# VIRUS-CELL INTERACTIONS
AND VIRAL ANTIMETABOLITES

# Seventh FEBS Meeting

**Volume 22**
VIRUS-CELL INTERACTIONS AND VIRAL ANTIMETABOLITES

**Volume 23**
FUNCTIONAL UNITS IN PROTEIN BIOSYNTHESIS

FEDERATION OF EUROPEAN BIOCHEMICAL SOCIETIES
SEVENTH MEETING, VARNA (BULGARIA), SEPTEMBER 1971

# VIRUS-CELL INTERACTIONS AND VIRAL ANTIMETABOLITES

Volume 22

*Edited by*

## D. SHUGAR

*Institute of Biochemistry and Biophysics,
Academy of Sciences; and
Department of Biophysics, University
of Warsaw, Warsaw, Poland*

 1972

ACADEMIC PRESS . London and New York

ACADEMIC PRESS INC. (LONDON) LTD.
24/28 Oval Road,
London NW1

*United States Edition published by*
ACADEMIC PRESS INC.
111 Fifth Avenue
New York, New York 10003

Copyright © 1972 by the Federation of European Biochemical Societies

*All Rights Reserved*
No part of this book may be reproduced in any form by photostat, microfilm, or any other means, without written permission from the publishers

Library of Congress Catalog Card Number: 75-189831
ISBN: 0-12-640866-1

Printed in Great Britain by
William Clowes & Sons Limited
London, Colchester and Beccles

# Preface

The development of appropriate vaccines has resulted, in many countries, in the virtual elimination of such viral diseases as polio and smallpox. But the vital problem of combatting viral infections already established in the organism is still very much with us. At the moment we can only take consolation in the fact that current research efforts appear to hold out the promise of at least some success in this direction in the near future. This was one of the reasons which dictated the selection of the subject of this symposium at the 6th FEBS Meeting in Madrid, two years previously, i.e. even prior to the exciting new advances in the field of the RNA oncogenic viruses, with their striking confirmation of the earlier proposals of Temin regarding the existence and role of reverse transcriptase in the mode of action of oncornaviruses. These developments, apart from the new perspectives they have opened up in the search for effective agents against tumour and other viruses, have in turn had a profound impact on the whole of molecular biology, the full implications of which cannot as yet be assessed.

It is obviously not feasible in a Symposium such as this to cover more than several facets of the multitude of specialized methods of approach to the study of virus-cell interactions and the problem of viral diseases. But, as will be seen from the contents of this volume, an attempt has been made to gather a sufficiently varied number of participants involved in different disciplines to provide a reasonably broad outline of progress both in basic research and in the development and clinical evaluation of the newer anti-viral agents. It is our hope that, as in the past, the presentations will also prove reasonably informative to the non-specialist.

The division of this Symposium into two distinct sections, one on Virus-Cell Interactions, the other on Viral Antimetabolites, was clearly dictated by reasons of convenience. It is unrealistic to consider a viral antimetabolite as an isolated entity and, in fact, only an adequate understanding of the various phases of virus-cell interactions, and the accompanying biochemical processes, can be expected to render possible a more rational approach to the design of therapeutically effective antiviral agents, as compared to the laborious screening techniques still widely employed, but which nonetheless have provided us with such promising drugs as 5-iododeoxyuridine, the rifamycins, the arabinosyl nucleosides, amantadine, etc.

These introductory remarks would be incomplete without an expression of appreciation to the Bulgarian Biochemical Society for the hospitality extended to the Symposium participants, and to the co-organizers, Dr. D. Naskkov and Dr. E. Golovinsky, for their untiring efforts prior to, and during the course of, the meetings.

*January 1972* D. SHUGAR

# Contents

| | |
|---|---|
| PREFACE | v |
| INTERACTIONS BETWEEN HOST GENOME AND ONCORNAVIRUSES IN ONCOGENESIS. By P. Bentvelzen | 1 |
| ANTIGENIC CHANGES IN CELLS INFECTED BY RNA TUMOUR VIRUSES. By G. Pasternak | 15 |
| EPSTEIN-BARR VIRUS-INDUCED ANTIGENS AND MACROMOLECULAR SYNTHESIS IN HUMAN, EBV-INFECTED LYMPHOBLASTOID CELL LINES. By I. Ernberg | 27 |
| EXPRESSION OF THE DNA-TUMOUR VIRUS GENOME: RELATIONSHIP TO DNA REPLICATION. By G. Sauer, C. Collins and H. Fischer | 33 |
| PROTEIN SYNTHESIS IN INTERFERON-TREATED AND VIRUS-INFECTED CELLS. By I. M. Kerr, P. Dobos, E. M. Martin, D. H. Metz and M. Esteban | 45 |
| NUCLEIC ACIDS AS INTERFERON INDUCERS. By E. De Clercq | 65 |
| STUDIES WITH TEMPERATURE-SENSITIVE MUTANTS OF ADENOVIRUS TYPE 5. By N. M. Wilkie, S. Ustacelebi and J. F. Williams | 87 |
| SPECIFIC RIBOSOMES: COMPONENTS OF AN ONCOGENIC RNA VIRUS. By J. Říman, J. Korb and A. Michlová | 99 |
| REVERSE TRANSCRIPTASE IN ONCOGENIC RNA VIRUSES. By S. Speigelman and J. Schlom | 115 |
| VIRAL AND HOST CELL INTERACTIONS WITH 5-IODO-2′-DEOXYURIDINE (IDOXURIDINE). By W. H. Prusoff | 135 |
| ANTIVIRAL ACTION OF OXIDIZED POLYAMINES. By U. Bachrach | 149 |
| A MOLECULAR ORBITAL APPROACH TO THE STUDY OF SOME STAGES OF PURINE METABOLIC PATHWAYS. By J. Kaneti and E. Golovinsky | 163 |
| MODE OF ACTION OF RIFAMYCIN AND AMINOPIPERAZINE DERIVATIVES ON ANIMAL VIRUSES AND CELLS. By Lise Thiry and G. Lancini | 177 |

ANKYLATED PYRIMIDINE NUCLEOSIDES AND (POLY)NUCLEO-
TIDES AS POTENTIAL ANTIVIRAL AGENTS. By D. Shugar  193
AUTHOR INDEX . . . . . . . . . . . 209
SUBJECT INDEX . . . . . . . . . . . 225

# Interactions Between Host Genome and Oncornaviruses in Oncogenesis

P. BENTVELZEN

*Radiobiological Institute TNO, Lange Kleiweg 151,
Rijswijk (Z.H.), The Netherlands*

Tumours can be induced in animals by various factors such as ionizing radiation, a great variety of chemical compounds, viruses and so on. But the most interesting are those tumours which arise spontaneously. Many inbred strains of laboratory animals have been developed which display great differences in incidences of spontaneous tumours in the various organs. This initially led to the concept that such tumours are hereditary. Indeed there are a few cases of clear-cut inheritance of a neoplastic disease such as pulmonary tumours [12], ovarian tumours [56] and mammary carcinomas in mice [4, 47] and renal adenomas in rats [21], but in most cases the genetic situation proves to be far more complex [29]. In view of the inducing factors mentioned above, the idea of cancer as a genetic disease has been given up in general, although it is recognized that host genetic factors will play some role in either spontaneous or induced development of a tumour.

From some spontaneous tumours oncogenic viruses have been retrieved in which RNA-containing viruses are prevalent. It must be emphasized that the list of "natural" tumour viruses is small. Well established oncogenic RNA viruses (oncornaviruses) are the agents of leukaemia in chickens, mice, cavias and cats, of sarcoma in chickens, mice and cats, and of mammary carcinoma in mice. In contrast to the oncogenic DNA viruses, these agents constitute a homogeneous group of viruses with regard to structural, biochemical and biological characteristics [49]. The virions are spheres about 100 nm in diameter, having a spherical internal electron-dense structure located at a restricted site, and lipid-rich outer membranes, which are formed at the cell membrane from which the virions are liberated by a budding process. Their genome consists mainly of a large single-stranded RNA molecule ($10^7$ daltons), which presumably consists of four or five subunits held together by hydrogen bonds. They contain a specific RNA-dependent DNA polymerase and other enzymes associated with the production of a double-stranded DNA using viral RNA as initial template. The oncornaviruses are oncogenic although not necessarily exclusive to their natural hosts, in contrast to several oncogenic DNA viruses. They are in general not cytopathic, insofar as neoplastic conversion is not regarded as pathologic to the cell itself. The transmission of these viruses is mainly vertical, often prenatally and then in close association with the host genome. For a review of this group of viruses see not only Nowinski et al. [49], but also Vigier [65] and Montagnier [45].

Virus particles which have an appearance and biochemical properties similar to these definite cancer viruses have been found in several other vertebrate species including man, but proof is lacking for oncogenic activity in their hosts. Oncornavirus particles have also been observed in pulmonary adenomas by Rabotti [55], in hepatomas by Maca et al. [42] and in tumours of pituitary and adrenal glands in mice by Mitchell et al. [44]. It is very premature to assume that these particles would be aetiologically involved in the tumours where they have been seen. Murine leukaemia viruses, which have the same appearance, replicate in many different tissues such as the mammary gland, but we have never found any evidence for the induction of a mammary tumour by a leukaemia virus.

Oncornaviruses have been retrieved from some murine lymphomas and mammary tumours induced by radiation or carcinogenic chemicals by Kaplan [32] and Timmermans et al. [64]. However, the negative results in the search for tumour viruses in chemically induced neoplasms, including lymphomas and mammary tumours (also in my own laboratory), are too numerous to accept an all-viral theory of the origin of cancer as such. Nevertheless we are of the opinion that studies on interactions between host genome and oncornaviruses are highly relevant to the problem of spontaneous development of tumours.

Genetics of neoplasia has to be considered at two separate levels: (1) genetic susceptibility of the host to develop tumours of a certain type, and (2) the cellular genetic changes underlying neoplastic transformation. On this basis we try to review the following relationships between host genome and oncornaviruses: (a) genetic susceptibility to an oncornavirus; (b) genetic transmission of an oncornavirus; (c) integration of an oncornavirus into the host genome; (d) cellular genetic changes under influence of an oncornavirus with regard to neoplastic transformation; (e) influence of epigenetic state of the host cell on oncornavirus functions including neoplastic transformation.

The first two relationships concern mainly host genetic factors, whereas the three others are at the cellular genetic level.

## GENETIC SUSCEPTIBILITY TO AN ONCORNAVIRUS

The first studies in this field, as far as we know, were made by Korteweg [33, 34] with the mouse mammary tumour virus. His results indicate that only a few genes control the difference between highly susceptible and highly resistant.

More detailed studies by Heston et al. [30] and Dux [18] are in accordance with this view. Susceptibility to murine leukaemia viruses often seems to be governed by a single gene [2, 37, 51], although more complex relationships also have been observed by Lilly [39]. The observation by Lilly [39] in the murine leukaemia system that genes can control susceptibility to only one strain of virus has also been made with the avian tumour viruses [17, 52, 66].

By means of mammary gland transplantation, Dux and Mühlbock [19,20] established that major susceptibility to the mouse mammary tumour virus is localized in the gland itself. In a few resistant strains humoral factors seem to play a role [48] and they are probably of an immunological nature [5]. An interesting observation is that genetic resistance to the virus is associated with a poor replication of the agent [5]; the rate of virion production corresponds well with susceptibility [9, 28].

Since, in the avian tumour virus system, genes control susceptibility to viruses with the same coat antigens [66], genetic resistance seems to be a block in an early phase of the infection. There is no difference in the rate of virus absorption between different genotypes [16, 54] suggesting that the block will be at the level of penetration or uncoating of the virus. Mutations, which affect haemopoiesis, also influence the response to leukaemia viruses. In addition Odaka and Matsukura [50] proved, by means of bone marrow transplantation, that susceptibility is expressed at the level of the haemopoietic cells. Genes, which control membrane components in mice (the so-called H-2 antigens), strongly influence susceptibility to leukaemia viruses [38]. These genes do not seem to facilitate penetration of the virus but influence later events related to the disease. An attractive hypothesis is that these membrane components have an influence on crucial cell surface alterations associated with neoplastic behaviour.

It seems unlikely that different gene-physiological systems would control susceptibility to the three oncornavirus groups. For instance, a more extensive search may demonstrate the existence of genes in birds which affect replication of a leukosis virus, and so on. As a tentative hypothesis we assume that in all three systems susceptibility to the oncogenic effect of an oncornavirus is achieved by separate genes for virus penetration, replication and noninterference with alterations of the cell membrane.

## GENETIC TRANSMISSION OF AN ONCORNAVIRUS

The possibility that tumour viruses can be part of the genetic make-up of a vertebrate organism was first suggested by Lwoff [40] in his classical review on lysogeny in bacteria. There is as yet no evidence for naturally occurring vertical transmission of oncogenic DNA viruses as in the case of several oncornaviruses [26, 31, 35, 46, 47].

Leukemia of the AKR mouse strain behaves as a hereditary trait in crosses with low-leukaemia mouse strains [15]. The hereditary nature of this disease was also obvious from the finding of Fekete and Otis [23] that AKR ova, transferred to low-leukaemic mouse strains, produce mice which subsequently become leukaemic. The discovery of a leukaemia virus in this mouse strain by Gross [24] was in apparent conflict with this postulate. Because introduction of this virus into other mouse strains leads only to milkborne transmission instead of transmission by the gametes, it was concluded that this leukaemia virus is transmitted as a genetic factor in its natural host, the AKR strain [26, 35].

The spontaneous release of virus is a feature which does not fit well into the accepted scheme if the presumed genetic transmission of the virus is compared with lysogeny. The recovery of leukaemia viruses from lymphomas induced by irradiation in otherwise low-leukaemic mouse strains [25, 36] better supports the resemblance between both situations as Lwoff [40, 41] had in mind. In this respect the retrieval of leukaemia viruses from chemically induced neoplasms [for review see Kaplan (32)] is important, since several carcinogenic drugs are also inducing agents in lysogenic bacteria [22,41].

We have extensively studied the possibility of genetic transmission of mouse mammary tumour viruses (MTV) for several virus variants and mouse strains [4, 6, 9, 11, 47]. The results can be summarized as follows: In five genetically very different mouse strains MTV-variants were observed which are vertically transmitted by the gametes. Introduction of an MTV-strain into a mouse strain, to which it is not indigenous, does not lead to gamete-borne transmission but only to transfer by the milk. There seems to be a very close relationship between the host genome and MTV-variants with regard to transmission by the germ cells. We may exclude the possibility that susceptibility genes could be involved in this phenomenon. The BALB/c mouse strain is more susceptible to MTV-L than the C3Hf strain, which is the natural host of this virus. Nevertheless only in the C3Hf strain is MTV-L transmitted by the sex cells.

This exclusive relationship between host genome and virus strain with regard to transmission by the gametes is easily explained by the transmission of these viruses as genetic factors of their natural host. We assumed that in one of the mouse chromosomes a DNA copy of the viral RNA is present, which under certain circumstances can be transcribed, giving rise to viral RNA and eventually complete virus particles (Fig. 1). Usually spontaneous virus release proves to be a recessive trait except in one case (the GR strain), where it is dominant. Genetic analysis proves this property to be controlled by a single gene. Every cell of the GR strain seems to contain viral activity, which indicates that information for the virus is part of the hereditary material of this mouse strain.

In those mouse strains which do not show spontaneous release of virus, the appearance of MTV virions or antigens can be induced by X-rays or urethan. An interesting phenomenon is that these antigens also appear in aged animals. The

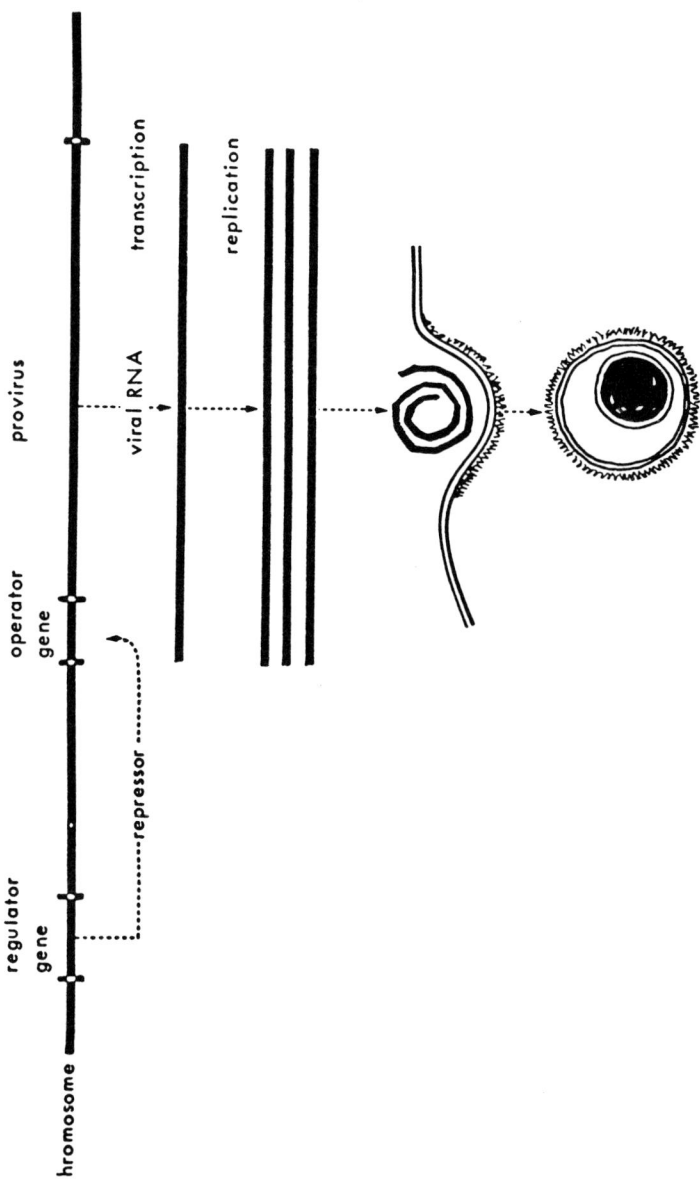

Figure 1. Scheme of genetic transmission of an oncornavirus.

time of appearance corresponds rather well with the susceptibility to spontaneous development of mammary tumours, suggesting that age-dependent switching on of an endogenous MTV is responsible for spontaneous carcinogenesis in the mammary gland.

The induction of MTV by carcinogens or ageing may be due to somatic mutations in controlling genes of the provirus. We favour, however, the hypothesis that this release is due to temporary derepression, an epigenetic event. In many cases we did not find complete virus particles, nor could we demonstrate infectivity of extracts from spontaneous tumours. Our technical procedures may be inadequate in that occasional virions have been overlooked and that too low doses of virus have been inoculated to find infectivity. The work of Hageman *et al.* [27] on the low-oncogenic variant of MTV is exemplary in this respect. However, one must remain aware of the possibility that in several mouse strains genetic entities are released, which are related to MTV but unable to produce complete virions and which are noninfectious. We advise calling such entities "viroids" as was suggested many years ago by Altenburg [1].

We observed a correlation between susceptibility to superinfection with the standard MTV strain and spontaneous release of an endogenous MTV. We have interpreted this as that a classical repressor, which prevents the transcription of the genetically transmitted provirus, also would interfere with the replication of a superinfecting virus. Further, we postulated that recessive mutations causing spontaneous release of virus were in the regulator gene, whereas the dominant lesion in the GR strain would be in the operator gene. In accordance with this whole concept is our observation that genes controlling susceptibility to superinfection are either linked or identical to genes controlling the release of endogenous virus. Furthermore the GR strain with its mutated operator gene is resistant to superinfection, indicating the presence of repressor. Also remarkable is the observed correlation between susceptibility to spontaneous and urethan-induced development of mammary tumours. We have likewise observed a similar correlation for pulmonary tumours in mice [10]. As in several instances the administration of urethan leads to the appearance of MTV-antigens, the observed correlation strongly pleads for an all-viral or, perhaps better stated, all-viroidal aetiology of mammary cancer in mice. Huebner and Todaro [31] have launched a similar general hypothesis concerning all modes of carcinogenesis. They suggest that certain determinants (oncogenes) of genetically transmitted C-type viruses (usually associated with leukaemia and sarcomas) are responsible for the development of most tumours. The many inbred mouse strains we have at our disposal display a great variation in tumour incidences of various organs. Therefore we are of the opinion that not one single oncogene is involved in the development of the various tumours. It cannot be completely excluded, however, that organ-specific expression genes control the switching on of one oncogene.

Huebner and Todaro [31] emphasized the frequent partial expression of the oncornaviruses. This may be reflected by independent appearance of internal or coat antigens of the virus or virus-coded cell-surface antigens and by the production of noninfectious entities, capable of transforming their host cell. In my laboratory we have failed to isolate a sarcoma virus from chemically transformed cells. However, my collaborator Sylvia Offers succeeded in retrieving such a virus when the sarcoma cells were infected with a leukaemia virus. Most likely the sarcoma virus resulted from the mixing (either phenotypically or by genetic recombination) of the sarcoma viroid with the infectious leukaemia virus.

An interesting example of partial expression of a provirus is the age-dependent release of MTV-O virions in BALB/c mice [27]. This mouse strain does not have a repressor causing immunity to superinfection with MTV. Nevertheless virions of the endogenous virus are not released in young adults but only in aged animals. Some virus-specific antigens are released throughout the life span. Obviously other control mechanisms can regulate the expression of some virus traits. It is remarkable that, upon passage of MTV-O in BALB/c mice, there is no interference with the production of complete virions.

On the basis of the repeatedly observed partial expression of oncornaviruses, Temin [61] developed the protovirus theory, which suggests that information for a whole virus is not necessarily present in the host genome. Furthermore he believes that virus release is a mutational event, whereas Huebner and Todaro [31] believe it to be epigenetic in nature. Techniques are as yet inadequate to discriminate between these postulates.

## INTEGRATION OF AN ONCORNAVIRUS INTO THE HOST GENOME

The finding of an RNA-dependent DNA polymerase in the virions of oncornaviruses by Baltimore [3], Temin and Mizutani [63], Spiegelman *et al.* [57] strongly substantiate the initially rather unorthodox theory of Temin [59] that a DNA copy is made from viral RNA. Since cellular DNA synthesis is needed before the virus can replicate [60], it seems logical to assume that this DNA copy is integrated into the host genome and will then serve as a template for synthesis of new viral RNA.

In the opinion of several authors the discovery of the reverse transcriptase warrants also the hypothesis of transmission of oncornaviruses as genetic factors of the host [49, 52, 57, 58]. In our opinion the provirus for vertical transmission (germinal provirus) does not need this enzyme for its continuity. Otherwise replacement of one germinal provirus by another following superinfection must be possible. We have never observed this in the MTV or murine leukaemia virus systems. If the RNA-dependent DNA polymerase were involved in vertical transmission, simple mendelian ratios, as found by Payne and Chubb [53], Bentvelzen [4], Bentvelzen and Daams [7] and Stockert et al. [58] would be impossible.

At first glance the hypothesis of the germinal provirus seems to be completely incompatible with the Temin postulate of the somatic provirus being produced after infection. In Fig. 2 a new hypothesis is presented, which reconciles both concepts. Ordinarily transcription of the germinal provirus is repressed. After temporary derepression induced by, for instance, radiation, viral RNA is released. Thereafter a new provirus is made and is inserted at a site less accessible to repressor molecules. This process accomplishes continuous production of viral RNA.

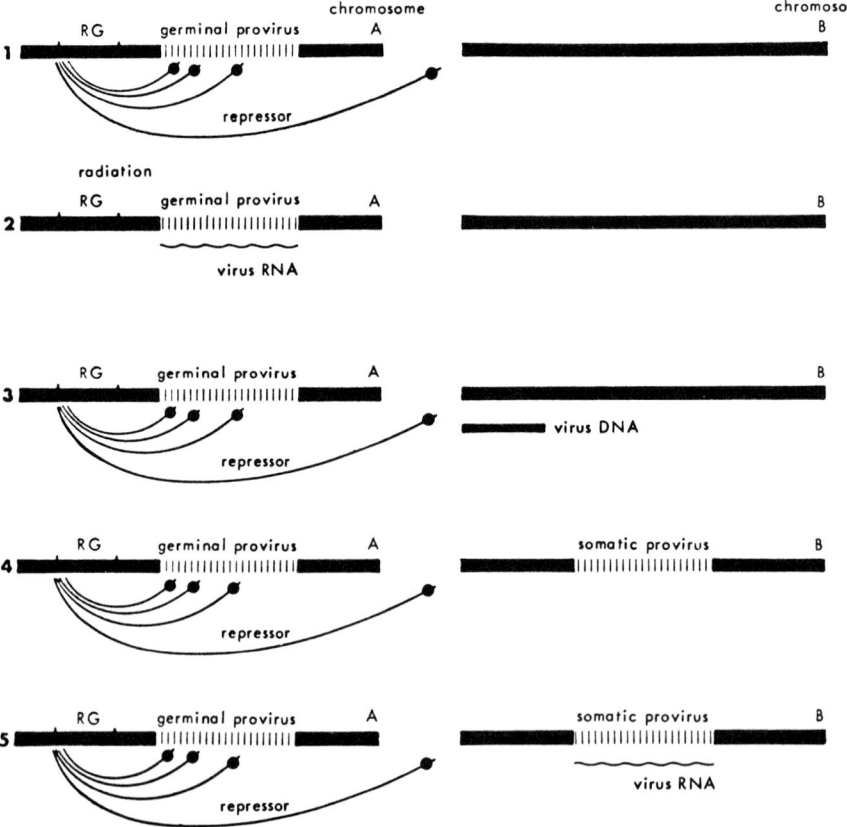

Figure 2. Combination of theories on germinal and somatic provirus: (1) repressed germinal provirus; (2) momentary derepression following radiation giving rise to transcription of germinal provirus; (3) restoration of repression; production of DNA copy of virus RNA; (4) insertion of DNA copy (somatic provirus) into another chromosome; (5) continuous transcription of somatic provirus.

## CELLULAR GENETIC CHANGES UNDER INFLUENCE OF AN ONCORNAVIRUS WITH REGARD TO NEOPLASTIC TRANSFORMATION

A source of heated discussions in cancer research is whether cancer is due to somatic mutations or epigenetic changes. One might argue that the establishment of a somatic provirus is a special form of somatic mutation, but in the case of endogenous oncornaviruses this would have been preceded by an epigenetic change.

Burdette and Yoon [14] found Rous sarcoma virus (RSV) to be mutagenic to *Drosophila*. It is very unlikely, however, that RSV would be oncogenic by the induction of point mutations. Macpherson [43] observed that the reversion of RSV-transformed hamster cells to normalcy is accompanied by the loss of the viral genome. The persistence of the viral genome seems to be necessary for maintenance of the neoplastic condition, which would not be the case with virus-induced mutations.

The correlation observed between virion production and susceptibility to carcinogenesis by MTV suggests that production of much viral RNA is necessary before neoplastic conversion can take place. An as yet unresolved problem is whether this conversion is the result of direct action of viral genes or the consequence of changes in expression of the host genome induced by viral products.

## INFLUENCE OF EPIGENETIC STATE OF THE HOST CELL ON ONCORNAVIRUS FUNCTIONS INCLUDING NEOPLASTIC TRANSFORMATION

Upon infection with the mammary tumour virus, infectivity can be retrieved from various tissues but only in secondary sex organs of the male and mammary glands of the female are complete virions produced. The virus has an oncogenic effect only in these female organs. In the lymphoid tissues some virus-specific antigens can be found which are not observed in the erythrocytes [9]. This demonstrates the considerable influence of the epigenetic state of the cell on various oncornavirus functions.

Murine leukaemia viruses can replicate in many different tissues but will transform only the haemopoietic ones. Our work with Rauscher leukaemia virus suggests a very subtle interplay between epigenetic events and the virus genome in the induction of erythroblastosis [13]. With antisera to the virus we could demonstrate the presence of the virus in haemopoietic stem cells. Various stimuli such as antigens, anti-platelet serum, which promote the proliferation of this stem cell, also enhance the leukaemic response. The kinetics of both processes closely resemble one other. The radiosensitivity of the stem cell parallels the effect of pre-irradiation on the leukaemic response to the virus. All these data suggest that the leukaemic process finds its origin in the stem cell. However, virus-infected stem cells are capable of normal functions: they can differentiate into thrombocytes, granulocytes or lymphocytes. It seems that the differentiation stimulus into the erythroid direction leads to tumorous derailment.

## CONCLUDING REMARKS

The following interactions between host genome and oncornaviruses have been discussed:

Genetic susceptibility to oncornaviruses is achieved by a low production of antibodies to the virus, a good penetration of the virus into the cell and subsequent uncoating, a good replication of the viral genome and tolerance to cell membrane alterations leading to neoplastic conversion.

Genetic vertical transmission of oncornaviruses is explained by the continuous presence of a DNA copy of the viral genome in one of the host chromosomes. Transcription of this germinal provirus takes place under the influence of germinal mutations in the controlling genes, after irradiation or treatment with carcinogenic drugs or ageing. Often only partial expression of the provirus is found.

After infection, oncornaviruses seem to make a DNA copy of their RNA, which becomes integrated into the cellular DNA. In the case of endogenous oncornaviruses the establishment of such a somatic provirus might be the escape from the repressor associated with the germinal provirus.

The persistence of the viral genome is needed for the maintenance of the neoplastic condition. The cell has to be in a certain epigenetic state before the virus can exert its oncogenic action.

It is not yet known whether neoplastic transformation is the consequence of direct viral action or of changes in the expression of the cellular genome induced by some viral products.

## REFERENCES

1. Altenburg, E. (1946). *Am. Naturalist* **80**, 559.
2. Axelrad, A. A. (1966). *Nat. Cancer Inst. Monogr.* **22**, 619.
3. Baltimore, D. (1970). *Nature Lond.* **226**, 1209.
4. Bentvelzen, P. (1968a). Genetic control of the vertical transmission of the Mühlbock mammary tumour virus in the GR mouse strain. Hollandia, Amsterdam.
5. Bentvelzen, P. (1968b). *J. Nat. Cancer Inst.* **41**, 757.
6. Bentvelzen, P. *In* "RNA Viruses and Host Genome in Oncogenesis" (P. Emmelot and P. Bentvelzen, eds), North Holland, Amsterdam (in press).
7. Bentvelzen, P. and Daams, J. H. (1969). *J. Nat. Cancer Inst.* **43**, 1025.

8. Bentvelzen, P. and Daams, J. H. (1970). *Europ. J. Cancer* **6**, 273.
9. Bentvelzen, P., Daams, J. H., Hageman, P. and Calafat, J. (1970). *Proc. Nat. Acad. Sci. U.S.A.* **67**, 377.
10. Bentvelzen, P. and Szalay, G. (1966). *In* "Lung Tumours in Animals" (L. Severi, ed.), pp. 835-844, University of Perugia, Italy.
11. Bentvelzen P., Timmermans, A. Daams, J. H. and Gugten, A. v. d. (1968). *Bibl. Haematol.* **31**, 101.
12. Bittner, J. J. (1938). *Public Health Rep. U.S.* **53**, 2197.
13. Brommer, E. J. P. and Bentvelzen, P. *In* "Proceedings of Vth Symp. Comparative Leukemia Research," Padova, in press, Karger, Basel.
14. Burdette, W. J. and Yoon, J. S. (1967). *Science, N.Y.* **155**, 340.
15. Cole, R. K. and Furth, J. (1941). *Cancer Res.* **1**, 957.
16. Crittenden, L. B. (1968). *J. Nat. Cancer Inst.* **41**, 145.
17. Crittenden, L. B., Stone, H. A., Reamer, R. H. and Okazaki, W. (1967). *J. Virol.* **1**, 898.
18. Dux, A. *In* "RNA Viruses and Host Genome in Oncogenesis" (P. Emmelot and P. Bentvelzen, eds), in press, North Holland, Amsterdam.
19. Dux, A. and Mühlbock, O. (1966). *Int. J. Cancer* **1**, 5.
20. Dux, A. and Mühlbock, O. (1968). *J. Nat. Cancer Inst.* **40**, 1309.
21. Eker, R. and Mossige, J. (1961). *Nature, Lond.* **189**, 858.
22. Epstein, S. S. and Saporoschetz (1968). *Experientia* **24**, 1245.
23. Fekete, E. and Otis, H. K. (1954). *Cancer Res.* **14**, 445.
24. Gross, L. (1951). *Proc. Soc. exp. Biol. Med.* **76**, 27.
25. Gross, L. (1958). *Acta Haematol.* **19**, 361.
26. Gross, L. (1961). "Oncogenic Viruses", Pergamon Press, New York.
27. Hageman, P., Calafat, J. and Daams, J. H. *In* "RNA Viruses and Host Genome in Oncogenesis" (P. Emmelot and P. Bentvelzen, eds), in press, North Holland, Amsterdam.
28. Hairstone, M. A., Sheffield, J. B. and Moore, D. H. (1964). *J. Nat. Cancer Inst.* **33**, 825.
29. Heston, W. E. (1965). *Cancer Res.* **25**, 1320.
30. Heston, W. E., Deringer, M. K. and Dunn, T. B. (1956). *J. Nat. Cancer Inst.* **16**, 1309.
31. Huebner, R. J. and Todaro, G. J. (1969). *Proc. Nat. Acad. Sci. U.S.A.* **64**, 1087.
32. Kaplan, H. S. (1967). *Cancer Res.* **27**, 1325.
33. Korteweg, R. (1936). Mitteilungen IIes Internationales Kongress Krebsforschung, Madrid, pp. 151-153.
34. Korteweg, R. (1940). *Acta Unio Internat. Contra Cancrum* **5**, 78.
35. Law, L. W. (1966). *Nat. Cancer Inst. Monogr.* **22**, 267.
36. Lieberman, M. and Kaplan, H. S. (1959), *Science, N.Y.* **130**, 387.
37. Lilly, F. (1966). *Nat. Cancer Inst. Monogr.* **22**, 631.
38. Lilly, F. (1968). *J. exp. Med.* **127**, 465.
39. Lilly, F. (1970). *J. Nat. Cancer Inst.* **45**, 163.
40. Lwoff, A. (1953). *Bacteriol. Rev.* **17**, 269.
41. Lwoff, A. (1960). *Cancer Res.* **20**, 820.
42. Maca, R. A., Heine, U. and Manaker, R. A. (1970). *Arch. Geschwulstf.* **36**, 213.
43. Macpherson, I. A. (1965). *Science, N.Y.* **148**, 1731.
44. Mitchell, W. M., Moses, H. L. and Orth, D. N. (1971). *Nature New Biol.* **231**, 99.

45. Montagnier, L. *In* "RNA Viruses and Host Genome in Oncogenesis". (P. Emmelot and P. Bentvelzen, eds), in press. North Holland, Amsterdam.
46. Moore, D. H. (1963). *Nature, Lond.* **198**, 429.
47. Mühlbock, O. and Bentvelzen, P. (1968). *In* "Perspectives in Virology" (M. Pollard, ed), Vol. VI, pp. 75-85, Academic Press, New York and London.
48. Nandi, S. (1967). *In* "Carcinogenesis, a broad critique" pp. 295-314, Williams and Wilkins, Baltimore.
49. Nowinski, R. C., Old, L. J., Sarkar, N. H. and Moore, D. H. (1970). *Virology* **42**, 1152.
50. Odaka, T. and Matsukura, M. (1969). *J. Virol.* **4**, 837.
51. Odaka, T. and Yamamoto, T. (1962). *Jap. J. exp. Med.* **32**, 405.
52. Payne, L. N. *In* "RNA Viruses and Host Genome in Oncogenesis". (P. Emmelot and P. Bentvelzen, eds), in press, North Holland, Amsterdam.
53. Payne, L. N. and Chubb, R. C. (1968). *J. gen. Virol.* **3**, 379.
54. Piraino, F. (1967). *Virology* **32**, 700.
55. Rabotti, G. F. (1966). *In* "Lung Tumours in Animals" (L. Severi, ed.), pp. 239-256, Univ. of Perugia, Italy.
56. Russell, E. S. and Fekete, E. (1958). *J. Nat. Cancer Inst.* **21**, 365.
57. Spiegelman, S., Burny, A., Das, M. R., Keydar, J., Schlom, J., Travnicek, M. and Watson, K. (1970). *Nature, Lond.* **227**, 563.
58. Stockert, E., Old, L. J. and Boyse, E. A. (1971). *J. exp. Med.* **133**, 1334.
59. Temin, H. M. (1964). *Nat. Cancer Inst. Monogr.* **17**, 557.
60. Temin, H. M. (1968). *Cancer Res.* **28**, 1835.
61. Temin, H. M. (1970). *J. Nat. Cancer Inst.* **46**, III.
62. Temin, H. M. *In* "RNA Viruses and Host Genome in Oncogenesis" (P. Emmelot and P. Bentvelzen, eds) in press, North Holland, Amsterdam.
63. Temin, H. M. and Mizutani, S. (1970), *Nature, Lond.* **226**, 1211.
64. Timmermans, A Bentvelzen, P., Hageman, P. C. and Calafat, J. (1969). *J. gen Virol.* **4**, 377.
65. Vigier, P. (1970) *In* "Progress in Medical Virology" (J. L. Melnick, ed.), pp. 240-283, Karger, Basel.
66. Vogt, P. K. and Ishizaki, R. (1966). *In* "Viruses Inducing Cancer" (W. J. Burdette, ed.), pp. 69-90. Univ. of Utah Press, Salt Lake City, Utah, U.S.A.

## DISCUSSION

**W. H. Prusoff:** Is there any relationship between the repressor content of "pro-virus" and the interferon content of the cell?

**P. Bentvelzen:** A few studies on the rate of interferon production in several mouse strains indicated that some strains, which are "repressed" in our system, do not produce interferon. It should be borne in mind that our mouse strains, which are not "repressed" for the mammary tumour virus, are "repressed" for the leukaemia virus, and *vice versa*. Consequently there appears to exist a specific inhibition of release of virus which is not explicable on the basis of interferon content.

**G. Pasternak:** What is the evidence that replication of oncornaviruses does not occur via an RNA minus strand?

**P. Bentvelzen:** The provirus theory has not yet been proven. The presence within the viron of the RNA-dependent DNA polymerase does, however, strongly support it. Inhibition of DNA synthesis, as well as actinomycin D, both prevent virus reproduction; this is most readily interpreted by the production of a DNA copy which serves as template for the synthesis of new viral RNA. The necessity of mitosis to promote virus production, as found by Temin, suggests that the DNA copy (provirus) must be integrated in the cellular genetic apparatus. There are two reports that double-stranded oncornaviral RNA can be found *in vivo*, but a number of laboratories have failed to confirm this. There are now several reports on the presence of a DNA copy in infected cells. I do not want to be dogmatic about the possibility that double-stranded RNA is not formed, but I believe it to be a dead end, i.e. that it will not play any role in virus replication.

# Antigenic Changes in Cells Infected by RNA Tumour Viruses

G. PASTERNAK

*Institute of Cancer Research,
Department of Immunobiology,
Berlin-Buch, G.D.R.*

RNA tumour viruses, such as the murine and avian leukaemia viruses, produce neoplasia in their indigenous hosts. Naturally, viruses or viral information are transmitted vertically from the parental host to the progeny. In this way, leukaemia may develop spontaneously. Artificial infection of susceptible hosts is likewise possible. It induces the disease after several weeks or months [2].

Leukaemia development is dependent on genetical factors of the host as well as on environmental conditions. It has been shown that mice of the H-$2^k$ genotype are most susceptible to infection and leukaemia induction by the Gross virus while the H-$2^b$ genotype seems to be relatively resistant [12]. Chemical carcinogens or X-rays apparently activate the viruses which are widely distributed among different mouse and chicken strains. Even resistant hosts may develop leukaemia when exposed to chemical or physical carcinogens. Cell-free extracts prepared from leukaemic tissue are infective and produce leukaemia. This has been extensively studied in mice by Lieberman and Kaplan [11], Irino et al. [10], Haran-Ghera [6].

Regardless of the activating factors, leukaemia development is associated with the appearance of new cell-bound antigens which cannot be detected in normal non-infected hosts [17]. These antigens render the malignant cells foreign to the host and induce an immune response. However, if the tumour is already established, it grows progressively.

The biological functions of the tumour-associated antigens are almost unknown. There are several data available now showing the various antigenic specificities in viral tumours and their distribution in different tissues. Even chemical analyses of certain antigens have been made. The interaction of the antigen-bearing cells with the immune system of the primary host, however, is still a matter of speculation.

In the following, the present state on the immunology of RNA virus-induced tumours will be summarized. The contribution is divided into four parts: (1) antigens of cells transformed by RNA tumour viruses; (2) antigenic changes of non-malignant cells infected with an RNA tumour virus; (3) antigenic changes of cells infected after transformation to malignancy; and (4) expression of viral antigens by activation.

Since the basic findings on the murine and avian viral tumours do not differ considerably, the following will be restricted to the immunology of murine leukaemia.

## ANTIGENS OF CELLS TRANSFORMED BY RNA TUMOUR VIRUSES

Mouse leukaemia cells are rejected in syngeneic hosts if the animals had been pre-immunized with X-ray-killed leukaemia cells, leukaemic extracts, or virus preparations [17]. Resistance to transplantation is specific since normal cells do not immunize against malignant cells and tumours of different aetiology grow progressively in leukaemia-resistant hosts. Resistance can be passively transferred to syngeneic recipients by immune cells or sera. Several *in vitro* techniques have been also developed to demonstrate the cellular and humoral reactions against the tumour-associated specific antigens (Table 1).

**Table 1.** Methods for the demonstration of virus-specified membrane-associated antigens of RNA viral tumors

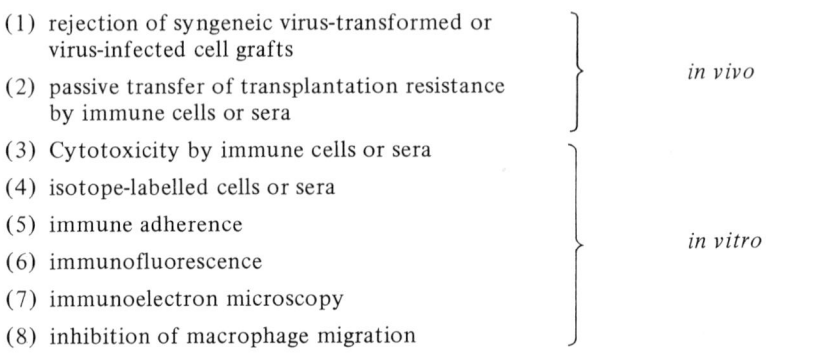

| | |
|---|---|
| (1) rejection of syngeneic virus-transformed or virus-infected cell grafts | |
| (2) passive transfer of transplantation resistance by immune cells or sera | *in vivo* |
| (3) Cytotoxicity by immune cells or sera | |
| (4) isotope-labelled cells or sera | |
| (5) immune adherence | |
| (6) immunofluorescence | *in vitro* |
| (7) immunoelectron microscopy | |
| (8) inhibition of macrophage migration | |

Leukaemias induced by the same or a closely related virus show cross-reactions. On the basis of the cross-experiments among leukaemias induced by different laboratory strains of virus, at least two groups of leukaemia can be distinguished: the Friend, Moloney, Rauscher and Graffi (FMRGr) group and the Gross (G) group which includes leukaemias induced by the so-called wild type viruses. Cross-reactions are absent between the FMRGr and G groups of leukaemia [15, 20]. The same antigenic specificity seems to be present on the virus, since the neutralizing activity of mouse immune sera shows a similar pattern of cross-reactivity.

Antigens responsible for graft rejection and cytotoxicity are those which are present on the cell membrane. In the case of viral leukaemia these antigens are virus-specified. They are termed virus-specified membrane-associated (MA) antigens. With respect to their nature there are three possibilities:

(1) The virus-specified MA antigens are viral envelope antigens which are present as an integral part of the cell membrane. Leukaemia cells are permissive in that they permanently release infectious virus by budding from the cell surface. Thus, antigenicity of the cell can be fully explained by the presence of viral antigens which may precede the virus release.
(2) A virus-induced new cellular antigen which does not show any antigenic relation to viral antigens is formed in transformed cells.
(3) Both types of antigen are present and are responsible for graft rejection and cytotoxicity There are a number of findings supporting the latter hypothesis [16, 25, 14, 1]. From these results, the virus-specified MA antigens consist of both viral and virus-induced new cellular antigens (Fig. 1).

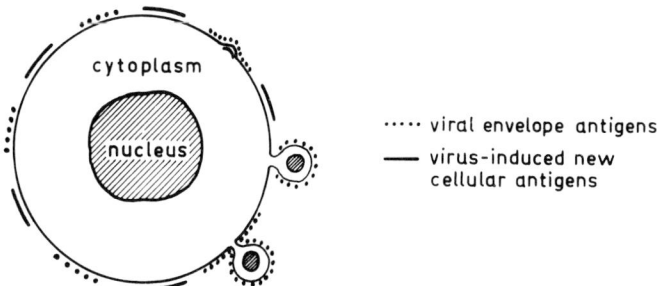

**Figure 1.** Virus release and virus-specified membrane-associated antigens of leukaemia cells as indicated by immunoelectron microscopy (a model). C-type viruses are released by budding from the cell membrane. During this process viral envelope antigens are present as an integral part of the membrane. Immunoelectron microscopy using ferritin-labelled antibodies shows that the sera label budding particles as well as areas of the membrane which do not show any morphological evidence of virus formation.

It is assumed that part of the surface antigens is a virus-induced new cellular antigen. Viral envelope antigens appearing at the sites of budding seem to occupy a larger membrane area around the particle. However, there are photographs in which labelling is strictly limited to the budding area. The antigens show a patchy surface distribution.

During virus release normal constituents of the cell membrane are apparently picked up by the viruses. Aoki* found in his immunoelectron microscopic studies the presence of $H\text{-}2^k$ antigen in 28.6% of the virions released from cells of this genotype. However, viruses budding from $H\text{-}2^b$ cells had the $\Theta$ antigen and not $H\text{-}2^b$. In our hands immunoelectron microscopy failed to detect normal

* Presented at the "International Symposium on Relationships between Tumor Antigens and Histocompatibility Systems," Paris, 26-28 February 1971.

cell antigens. Considering the possible distribution of surface antigens in virus-releasing areas, however, it seems very likely that viruses may pick up normal membrane antigens (Fig. 2).

Recent studies of several laboratories were concerned with another antigen of leukaemia viruses, the group-specific (**gs**) viral antigens [4, 5, 7, 24]. This subject became of particular interest when these antigens could be detected in various mouse tissues [9]. Even tissues from mice which do not develop leukaemia during their life-span were found to contain **gs** viral antigens.

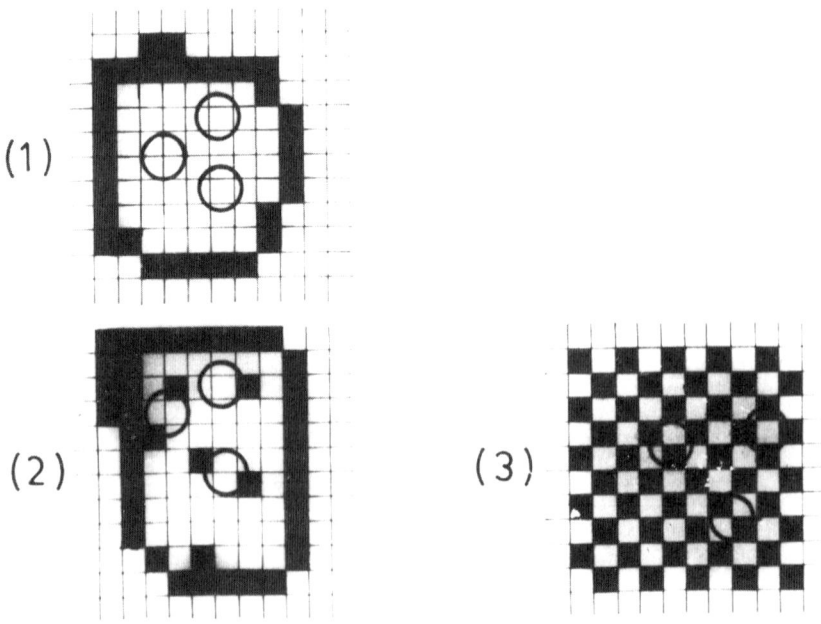

Figure 2. Hypothetical distribution of surface antigens in virus-releasing areas of the cell. Viruses which are released from the membrane may pick up normal cell antigens. However, this seems to be a relatively rare event. Virus release is thus supposed to take place at preformed areas of the membrane which mainly consist of viral envelope antigen. This conception is consistent with the models (1) and (2) but not with (3) in which a mosaic-like antigen structure would cause the uptake of normal antigens at a higher proportion of viruses. □ viral envelope antigen (type-specific); ■ antigens of the normal cell membrane.

While the envelope antigen of the virus is present on the surface, this type of antigen is located in the cytoplasm of the cell. It represents an internal viral antigen which is in the nucleoid component of the virus (Fig. 3).

Ouchterlony gel diffusion tests have shown that the **gs**-antigen is a complex of antigenic components of a generally similar nature. Two main types of **gs**-antigens can be distinguished in mammalian leukaemia viruses [23]. One is a **gs**-species antigen which is immunologically identical among leukaemia viruses of the same species and the other antigen is shared by the viruses of other species.

The latter one is termed the **gs**-interspecies antigen or the **gs**-3 antigen (Table 2). The antigen can be detected by sera from rats carrying large syngeneic grafts of leukaemic tissue. Sera of similar immunological specificity can be produced by immunizing rabbits with purified **gs**-antigen. Apparently the mouse is incapable of reacting against the **gs**-antigens. Hitherto there have been no reports on the

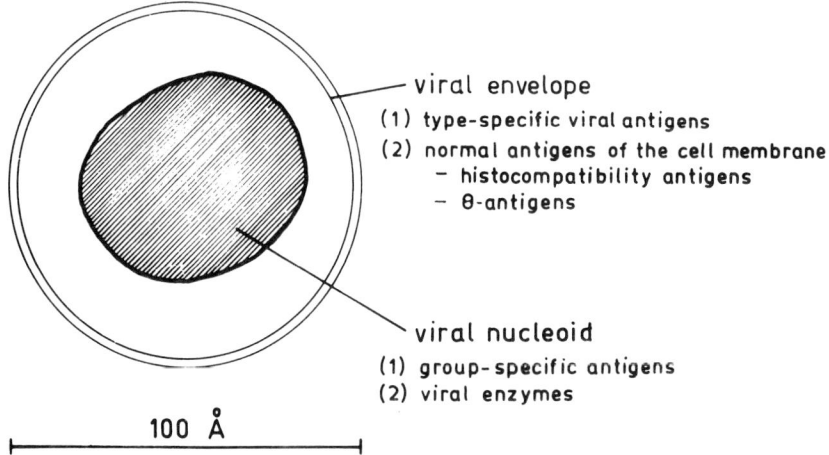

Figure 3. Antigens of C-type viruses.

demonstration of **gs**-antibodies in mice. It is assumed that mice are tolerant to this antigen. However, chicken form antibodies to the **gs**-antigen of avian leukosis viruses. In this case the original hypothesis on tolerance to **gs**-antigens has been disproven.

Table 2. Group-specific (gs) antigens of murine C-type viruses in virus or cell preparations of different species origin

|      | Mouse | Rat | Hamster | Cat | Bovine | Human | Avian |
|------|-------|-----|---------|-----|--------|-------|-------|
| gs 1 | +     | −   | −       | −   | −      | +?    | −     |
| gs 2 | +     | −   | −       | −   | −      | −     | −     |
| gs 3 | +     | +   | +       | +   | +?     | +?    | −     |

A very important fact is the presence of **gs**-antigens in the tissues of apparently non-infected animals.

As an example, embryonic tissue of wild mice may contain **gs**-antigens although leukaemia development in these mice is certainly a rare event (9). If the expression of this antigen depends on a viral gene the conclusion can be drawn that certain viral activities may function in an organism which is resistant to leukaemia.

The findings on the distribution of gs-antigens and other viral activities in chicken and mice were the basis of Huebner's oncogen and virogen hypothesis [8] which postulates that the information for malignancy, antigen expression, and virus production is already present in the cell. Additional factors are then responsible for the expression of all or only a few activities. This hypothesis represents the practical and theoretical perfection of previous conceptions of several investigators.

It is still not clear which of the gs-antigens detected in embryonic and other normal tissues is present in the virion. In the chicken system there are some indications of a structural difference between foetal and tumor-derived gs-antigens.*

Finally, it should be added that virus-producing cells have the viral envelope antigen in the cytoplasm. Most probably, part of the antigen is integrated in the membranes of the endoplasmic reticulum while another part is certainly present in soluble form.

## ANTIGENIC CHANGES OF NON-MALIGNANT CELLS INFECTED WITH AN RNA TUMOUR VIRUS

Non-malignant cells naturally or artificially infected with a mouse leukaemia virus may contain the same virus-specified antigens as malignant cells [17].

After virus infection the cells of spleen, lymph nodes, thymus, or bone-marrow are the targets that are potentially susceptible of being transformed to leukaemia cells. When derived from animals in the pre-leukaemic stage, shortly after infection of newborns, the cells are certainly still non-malignant.

In Graffi mouse leukaemia, which has a long latency period, MA leukaemia antigens are present in spleen cells already 14 days after infection [18]. In contrast, thymus, liver, and brain do not contain measurable amounts of antigen although the virus titre in these organs is about the same as the titre in spleen. The antigens are lacking in spleen cells of immune animals which had been infected as adults. However, the cells produce the virus. Apparently, virus production is not correlated with antigen expression. There are two possibilities which may explain this discrepancy.

(1) The concentration of viral envelope antigen on the cell surface is independent of the amount of virus released, i.e. a high concentration of the antigen such as in spleen is not necessarily paralleled by high virus production; or

* Rabotti presented this information at the Conference on "Biological Function of Tumor Specific Antigens Induced by Oncogenic Viruses" in Prague, 1971.

(2) certain cells acquire an additional antigen, the virus-induced new cellular antigen. Thus, spleen cells, which are the targets for myeloid Graffi leukaemia, have viral as well as virus-induced new antigens while the other cell types carry only viral antigen on the surface. Whether non-malignant cells have actually the same antigens as leukaemia cells is still an open question. Results of our experiments with spleen cells from animals infected as newborns are in favour of the second hypothesis. It is assumed that the presence of MA antigens at the cell surface does not render the cells neoplastic.

Studying the Gross virus-induced antigens, Wahren [27] was also able to show that lymphoid cells of two to nine months-old AKR mice which spontaneously develop leukaemia contain MA Gross leukaemia antigens. Furthermore, Breyere and Williams [3] presented evidence that normal skin of leukaemic mice possesses virus-specified MA antigens. Skin grafts from these animals were rejected in virus-immunized mice. The antigen involved has not been characterized.

## ANTIGENIC CHANGES OF CELLS INFECTED AFTER TRANSFORMATION TO MALIGNANCY

Infection of tumours unrelated to leukaemia as well as super-infection of leukaemia with another type of leukaemic virus produces antigenic conversion. Antigenic conversion is the acquisition of an MA antigen which is produced as a consequence of virus infection. The term "antigenic conversion" was introduced by Stück *et al.* [26] who artificially infected with the Rauscher virus a variety of transplanted ascites leukaemias and sarcomas lacking the FMRGr antigen.

Infection occurred after passaging the cells in infected animals. Some of the leukaemias tested were found to have acquired the FMRGr antigens after this procedure. Previous experiments by Pasternak and Pasternak [21, 22] have shown that tumours originally induced by a chemical carcinogen acquire MA leukaemia antigens if they had been infected with Graffi virus. Syngeneic cell inocula were rejected in Graffi virus-immunized animals. Indirect evidence was obtained from absorption experiments that the antigen involved is the viral envelope antigen. However, the experiment does not exclude the presence of low amounts of a virus-induced new cellular antigen on the cell surface (Fig. 4).

If a virus-producing leukaemia cell is superinfected with an unrelated leukaemia virus, then, theoretically at least, four different antigens are possible to be expressed, i.e. two viral envelope antigens and two virus-induced new cellular antigens.

Our experiments with Graffi virus-infected Gross leukaemia cells have shown, however, that interference of antigenic expression may occur. The superinfected cells do produce the virus and thus contain a viral antigen at the surface, but they are not sensitive to cytotoxic antibodies against Graffi leukaemia and to the rejection response of Graffi virus-immunized animals [13].

Probably due to the presence of Gross antigens and/or the production of Gross virus, the cells or some virus products suppress the expression of MA antigens of Graffi leukaemia, although the latter virus is produced. The Gross antigens interfere with the Graffi antigens. We call this phenomenon "incom-

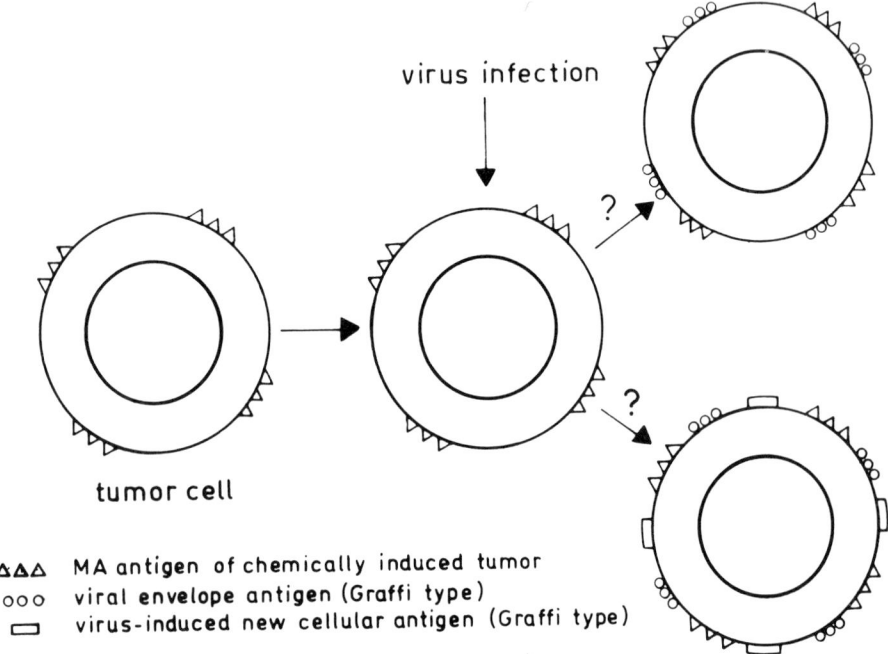

△△△ MA antigen of chemically induced tumor
ooo viral envelope antigen (Graffi type)
▭ virus-induced new cellular antigen (Graffi type)

Figure 4. Antigenic conversion of chemically induced tumour by infection with a mouse leukaemia virus. A methylcholanthrene-induced sarcoma carrying a specific TSTA acquires virus-specified antigens after *in vivo* infection with a leukaemia virus. The cells are permanently releasing virus and thus they contain viral envelope antigens on the surface. Presence of this viral antigen would be sufficient to explain the new antigenicity of the cells. However, the alternative cannot be excluded that a virus-induced new cellular antigen is present in addition to the viral envelope antigen. The original TSTA of the chemically induced tumour does not change after infection.

plete antigenic conversion" [20]. Concerning the question as to which antigen of the Graffi leukaemia type is involved, the same dilemma exists as in antigenically converted sarcoma cells or virus-producing normal cells. It is not known whether only one or two antigens are suppressed (Fig. 5). The term antigenic conversion can certainly also be applied to infected normal cells, which have been shown to acquire a new antigen.

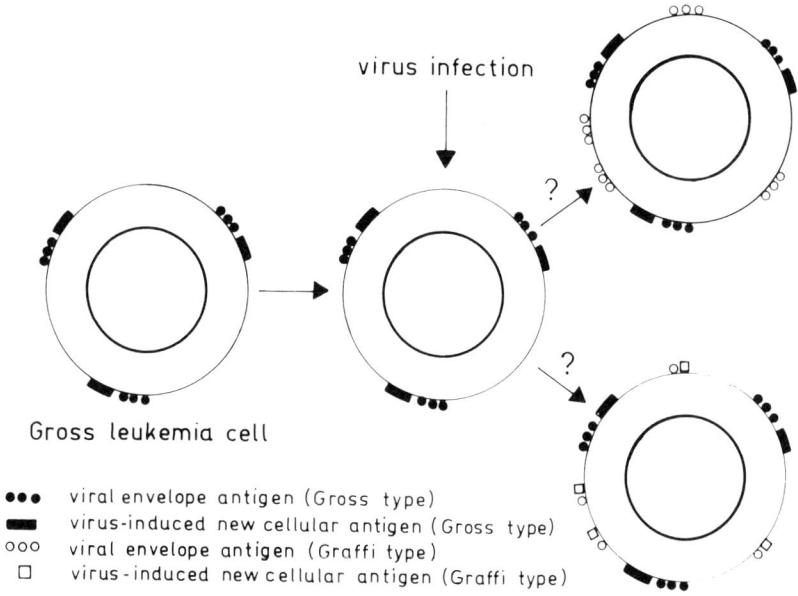

**Figure 5.** Antigenic conversion of Gross leukemia cells by infection with Graffi leukaemia virus. Four distinct virus-specified antigens are theoretically possible on the cell surface. However, experimental data show that expression of antigens induced by the superinfecting agent is suppressed. Only the immunofluorescence test indicates virus-specified antigens of the superinfecting agent. It is postulated that the new antigenicity of the cell is either due to viral antigens, which can be detected by absorption or to both viral antigens and virus-induced new cellular antigens. In this case, the new cellular antigen is below the concentration detectable.

## EXPRESSION OF VIRAL ANTIGENS BY ACTIVATION

Leukaemias induced in mice of strain XVII by the chemical carcinogen methyl nitrosourea (MNU) show expression of viral antigens [19]. Obviously, a virus is associated with the appearance of this type of leukaemia. Serological studies indicate that the MA leukaemia antigen is of the Gross type. Although the mouse strain used has a very low incidence of spontaneous leukaemia ($< 1\%$) the majority of chemically induced leukaemias contains the MA Gross leukaemia antigen. Apparently the chemical carcinogen produces leukaemia by activation of a virus which is present in all mice of the strain but which, under normal conditions, is incapable of being leukaemogenic. At the same time a large number of the chemically induced leukaemias exhibits gs-antigens. However, there are a few exceptions. It was found that some leukaemias may have the MA antigen in the absence of gs-antigens and vice versa. Even leukaemias having no detectable antigen developed after MNU (Table 3). Regardless of the antigenic make-up, the tumours had the same degree of malignancy and they were likewise transplantable to syngeneic hosts.

Table 3. Expression of viral activities in murine leukaemia produced by a chemical carcinogen (methyl nitrosourea)

| Type of leukaemia | Frequency of appearance (%) | MA antigens of Gross type | gs antigens of MuLV | Transplantability |
|---|---|---|---|---|
| I   | ~80 | + | + | + |
| II  | ~8  | + | − | + |
| III | ~8  | − | + | + |
| IV  | ~4  | − | − | + |

To summarize, virus-producing non-malignant, as well as transformed cells, may contain the same virus-specified MA antigens. MA antigens are the virus-induced new cellular antigens and antigens of the viral envelope which are an integral part of the membrane. Intracellularly localized viral antigens are the gs antigens of the viral nucleoid which are characteristic of the murine or avian leukaemia and sarcoma viruses. Gs antigens are also present in normal tissues of apparently non-infected animals.

Infection of malignant cells, as well as superinfection with an unrelated virus, may produce antigenic conversion. There are some indications that superinfection of a mouse leukaemia cell with a closely related leukaemia virus produces interference of antigenic expression. Expression of viral antigens has been also detected in chemically induced mouse leukaemia.

On the basis of the experimental data obtained, the hypothesis is proposed that expression of the MA antigens is not correlated with malignant transformation. Transformation may occur without expression of MA leukaemia antigens and antigen expression is possible in non-malignant cells. The question whether virus release parallels the expression of MA leukaemia antigens is not definitely solved. Furthermore, the data show that the expression of MA antigens, gs-antigens, and oncogenicity are independent entities.

REFERENCES

1. Aoki, T., Boyse, E. A., Old, L. J., de Harven, E., Hämmerling, U. and Wood, H. A. (1970). *Proc. nat. Acad. Sci. U.S.A.* **65**, 569.
2. Bierwolf, D., Fey, F., Graffi, A., Pasternak, G. and Schramm, T. (1968). *Curr. Topics Microbiol. Immunobiol.* **46**, 26.
3. Breyere, E. J. and Williams, L. B. (1964). *Science, N.Y.* **146**, 1055.
4. Geering, G., Old, L. J. and Boyse, E. A. (1966). *J. exp. Med.* **124**, 753.

5. Geering, G., Hardy, W. D., Old, L. J. and de Harven, E. (1968). *Virology* **36**, 678.
6. Haran-Ghera, N. (1967). *Proc. Soc. exp. Biol.* (*N.Y.*) **124**, 697.
7. Huebner, R. J. (1967). *Proc. nat. Acad. Sci. U.S.A.* **58**, 835.
8. Huebner, R. J. and Todaro, G. J. (1969). *Proc. nat. Acad. Sci. U.S.A.* **64**, 1087.
9. Huebner, R. J., Kelloff, G. J., Sarma, P. S., Lane, W. T., Turner, H. C., Gilden, R. V., Oroszlan, S., Meier, H., Myers, D. D. and Peters, R. L. (1970). *Proc. nat. Acad. Sci. U.S.A.* **67**, 366.
10. Irino, S., Ota, Z., Sezaki, T., Suzaki, M. and Hiraki, K. (1963). *Gann* **54**, 225.
11. Lieberman, M. and Kaplan, H. S. (1959). *Science N.Y.* **130**, 387.
12. Lilly, F. (1966). *Nat. Cancer Inst. Monogr.* **22**, 631.
13. Micheel, B. and Pasternak, G. (1968). *Int. J. Cancer* **3**, 603.
14. Oboshi, S., Hakura, K. and Maruyama, K. (1967). *Gann* **58**, 367.
15. Old, L. J., Boyse, E. A. and Stockert, E. (1964). *Nature, Lond.* **201**, 777.
16. Pasternak, G. (1967). *Nature, Lond.* **214**, 1364.
17. Pasternak, G. (1969). *Adv. Cancer Res.* **12**, 1.
18. Pasternak, G. and Pasternak, L. (1967). *J. nat. Cancer Inst.* **38**, 157.
19. Pasternak, G. and Pasternak, L. (1971). *Boll. Ist. Sieroter. Milan.* **50**, 192.
20. Pasternak, G., Pasternak, L. and Micheel, B. (1971). *In* "RNA Viruses and Host Genome in Oncogenesis", Symp. on the occasion of the 65th birthday of Prof. O. Mühlbock, Amsterdam, 1971, in press, North-Holland Publishing Comp., Amsterdam.
21. Pasternak, L. and Pasternak, G. (1968). *Arch. Geschwulstforsch.* **31**, 243.
22. Pasternak, L. and Pasternak, G. (1968). *Arch. Geschwulstforsch.* **32**, 301.
23. Schäfer, W. and de Noronha, F. (1971). *J. Am. Vet. Med. Ass.* **158**, 1092.
24. Schäfer, W., Lange, J., Pister, L., Seifert, E., de Noronha, F. and Schmidt, F. W. (1970). *Z. Naturf.* **25b**, 1029.
25. Steeves, R. A. (1968). *Cancer Res.* **28**, 338.
26. Stück, B., Old, L. J. and Boyse, E. A. (1964). *Nature, Lond.* **202**, 1016.
27. Wahren, B. (1966). *Int. J. Cancer* **1**, 41.

DISCUSSION

**H. A. Blough**: What is the evidence for the synthesis of membrane-associated antigens in the endoplasmic reticulum? And are they the same as the antigens found in the cell membrane, particularly in terms of chain elongation of sugars (glucosyl transferase activity)?

**G. Pasternak**: The presence of membrane-associated antigens in the endoplasmic reticulum can be detected by adsorption of immune sera with membrane fractions. As far as I know, there are as yet no data available on the chemical composition of the virus-specified membrane-associated antigens.

**L. Thiry**: Recent studies have shown that transformed cells grown in the presence of dibutyril CAMP acquire some surface properties of normal cells such as contact inhibition. Have you any information on the nature of the membrane-associated antigens following growth in the presence of dibutyril CAMP?

**G. Pasternak**: There are certainly additional changes on the cell surface which are not virus-induced. In addition to the membrane-associated antigens, embryonic antigens have been detected on the surface of leukaemia cells (personal communication from Dr. Della Porta of Milan). I have no information on the changes occurring after treatment of the cells with cyclic AMP.

**O. P. van Diggelen**: Is the ATPase activity of AMV an integral part of the viral envelope, or is it merely associated in some looser fashion?

**G. Pasternak**: I think Dr. Říman could better answer this.

**J. Říman**: In all probability the ATPase is an integral part of the virus, since it is present only when virus has been grown on myeloblasts. Integration of cellular structures consequently is not a very rare event.

# Epstein-Barr Virus-induced Antigens and Macromolecular Synthesis in Human, EBV-infected Lymphoblastoid Cell Lines

## INGEMAR ERNBERG

*Department of Tumour Biology,*
*Karolinska Institutet, Stockholm, Sweden*

The Epstein-Barr virus [2] is a herpes type virus. It is the causative agent of infectious mononucleosis [8, 14]. Several herpes viruses are oncogenic in animals, e.g. herpes saimiri in monkeys and Marek's disease virus in chickens. It is possible that the Epstein-Barr virus is oncogenic in man. Many studies implicate a role of the virus in the aetiology of Burkitt's lymphoma and nasopharyngeal carcinoma [13]. Tumour biopsies from patients with these diseases can give rise to established cell lines of lymphoblastoid type. Such lines can also be obtained from blood cells of healthy donors. All cultured cell lines studied contain EBV-DNA-like DNA. The technique used was molecular hybridization [6]. Human lymphoblastoid cell lines are so far the only object in which the behaviour of the virus can be studied.

The molecular weight of the DNA of herpes viruses is around $10^8$ dalton, which gives them a potential of coding for about 200 medium sized proteins. Compared to this, the number of virus-associated antigens is still few. They are divided into three groups by the methods used: complement fixing antigens [1, 19], immunoprecipitating antigens [15] and antigens detected by immunofluorescence.

The lymphoblastoid cell lines are of two types with regard to the antigens: (1) those which continuously produce virus-associated antigens as detected by immunofluorescence and, (2) those which do not produce virus-associated antigens as detected by IF. The latter group was originally referred to as EBV-free. Recently it has been shown that these cells contain complement fixing antigens [16] and also, as mentioned, EBV-DNA-like DNA. The relationship between the antigens detected by the different methods is not clear.

So far there are three known antigens detected by immunofluorescence, of which two are divided into subgroups. The viral capsid antigen, VCA [6], was demonstrated in cells containing virus particles. Recently it was also shown, by immunoferritin labelling, that antibodies to VCA label naked, but not enveloped, particles [17]. The membrane antigen, MA [11], on the other hand, is present on enveloped but not naked particles and on the outer plasma membrane of cells derived from EBV-carrying culture lines. The early antigen, EA, has been found to appear in certain blastoid cell lines, negative for IF-positive EBV antigens, after EBV-infection [9]. Its possible presence in the virions has not been studied. The membrane antigen, MA, is divided into three subunits. All three appear at the same site on the cell and at the same time. Anti-MA positive sera contain antibodies against one or several subunits [18]. The early antigen consists of two known subunits, "restricted" (R) and "diffuse" (D) of different antigenic specificities [10].

Antibodies to the early antigen were best detected in the sera of patients with nasopharyngeal carcinoma (NPC). Many NPC sera, and BL sera as well, contain antibodies to MA, EA and VCA. Anti-EBV (VCA) positive IgG from healthy donors very rarely exhibit anti-EA activity. There is a good correlation between the anti-VCA and anti-MA reactivity in about 80% of human sera, although exceptional "discordant" sera show preferentially anti-MA or anti-VCA activity with low or negative titres in the other test. There are no known anti-EA positive, VCA negative sera. For the detection of antigens we now use direct immunofluorescence, conjugating representative reference sera with fluoroscein isothiocyanate or rhodamin isothiocyanate. In this way we can obtain double-stained cells and study them under appropriate filters for red and green fluorescence [12].

Many cell lines of the above-mentioned type 2, i.e. those that do not produce virus-associated antigens as detected by immunofluorescence, can be infected with EB-virus. Five minutes after infection, membrane fluorescence can already be seen on the cells. This fluorescence appears even if uv-inactivated virus is used or if the cells are pretreated with puromycin in doses completely inhibiting protein synthesis. This immediate membrane fluorescence disappears slowly with time. It can be washed away, at least in part. About 10 hours after infection a new peak of membrane fluorescence appears. This peak is not obtained in puromycin treated cultures or with uv-inactivated virus. As already mentioned the virus contains MA in the envelope. The first peak of MA may be due to adsorbed virus, whereas the latter represents a new protein synthesized by the cell and inserted into the cell membrane [3].

Parallel with the appearance of MA on the surface, intracellular EA appears in some cells. The number of EA-positive cells depends on the virus dose. In infected cells of the Raji line, EA is first seen in the nucleus and later spreads to the cytoplasm. The two subunits R and D appear simultaneously, as a rule. In rare instances VCA appears in EA-containing cells about 40 hours after infection. No infective virus is released, and the infection is aborted. If $^3$H-thymidine incorporation is inhibited by the addition of cytosine arabinoside or iododeoxyuridine, the course of the infection remains essentially unchanged. Puromycin prevents the appearance of EA or VCA.

The susceptibility of different cell lines to the virus, as measured by the appearance of intracellular antigens, is variable. Some lines are very sensitive to infection, like Raji and Daudi; whereas other cell lines, like 6410, are relatively resistant.

Cell lines of the type 1, referred to above, continuously produce virus-associated proteins, as detected by immunofluorescence in a minority of the cells. They have a variable number of MA-positive cells (0-70%), depending on culture conditions [20]. The number of VCA- and EA-positive cells vary as well. Under normal culture conditions there are about 2-5% EA-positive cells and less than 1% VCA-positive cells in several lines of this type. The types of cells present in these lines are thus antigen-negative cells, MA + EA − VCA − cells, MA + EA + VCA − cells, MA + EA + VCA + cells and a few MA − EA + VCA − cells. It seems that cells are continuously thrown off from the main cell line, characterized by a normally permissive virus-cell interaction, which enters into a more or less abortive viral cycle. Only one cell line, P3HR-1, proceeds regularly to the release of infectious virus in easily detectable amounts.

We have assessed the macromolecular synthesis in the different categories of cells by $^3$H-thymidine, $^3$H-uridine and $^3$H-phenylalanine incorporation, respectively. Fluorescence labelled smears were prepared as usual, dipped in photoemulsion and exposed for three to six days. The radioautography grains and fluorescence could be observed in parallel in the uv-microscope [4]. In infected Raji cells, EA-positive cells ceased to synthesize DNA about 20 hours after infection; 40-50 hours later RNA and protein synthesis were brought to a standstill as well, and the cells were obviously dying. Antigen-negative, or MA + EA − cells, continued their DNA, RNA and protein synthesis.

In EBV-antigen producing cell lines of type 1, analogous phenomenon could be observed. DNA-inhibitors, e.g. cytosine arabinoside, iododeoxyuridine and mitomycin C induced an accumulation of EA-positive cells. On the other hand, viral capsid antigen was only present when DNA synthesis was permitted. VCA-positive cells could be made to appear in ara-C inhibited cultures by washing the cells and adding deoxycytidine. The optimal effect was obtained if this was done 40-50 hours after addition of ara-C [5]. Under these conditions, the few viable EA-positive cells extended their protein synthesis about 10 hours, DNA-synthesis and VCA appeared. Subsequently, the MA + EA + VCA + cell turned off its macromolecular synthesis 10-15 hours after the appearance of VCA.

The presence of VCA inside a cell was reported to parallel the presence of virus particles [7]. Many VCA-positive cell lines, with P3HR-1 as the notable exception, rarely released demonstrable quantities of infectious virus. This indicates that the virus can be trapped in the VCA-positive cells or that incomplete virus is being released by many cell lines.

In the carrier cultures there is a continuous induction of infectious viral cycle in a few cells. This phenomenon is also seen in the presence of virus neutralizing serum. Furthermore the early antigen appears in a higher frequency when DNA-inhibitors are added. These inhibitors do not allow the production of infectious virus. The continuously appearing cells with late viral proteins are thus not the result of reinfection by released virus. Internal events must therefore lead to the synthesis of virus-associated proteins in some cells. Cloning experiments have shown [21] that all clones yield antigen-positive cells in proportions similar to those in the original culture. This indicates that all cells contain the EB-viral genome and have the same potentiality to express it, characteristic for the line. The analogy with lysogenic bacteria is striking.

MA-positive cells are of great interest. This is the only antigen present in biopsies from tumours in man. There is recent evidence (G. Klein, personal communication) that MA on EBV-carrying culture lines, and on BL biopsy cells, have identical or cross-reactive specificities. Tumour cells persisting in spite of therapy often become coated with IgG and these antibodies react specifically with MA on cultured cells. Membrane changes induced by viruses may play a role in transformation of cells. In carrier cultures, MA-positive cells exhibit a rate of DNA synthesis somewhat lower than in antigen-negative cells. The rate of protein and RNA synthesis is the same in MA-positive cells and antigen-negative cells. Although we have not yet been able to study directly whether MA-positive cells are able to divide, our findings make MA-positivity potentially compatible with continued cell life and proliferation.

This study was conducted under USPHS Contract No. NIH-69-2005 within the Special Virus Cancer Program of the National Cancer Institute, National Institutes of Health. Grants were also received from the Swedish Cancer Society.

## REFERENCES

1. Armstrong, D., Henle, G. and Henle, W. (1966). *J. Bact.* **1**, 1257-1262.
2. Epstein, M. A., Achong, B. G. and Barr, Y. M. (1964). *Lancet* **I**, 702-703.
3. Gergely, L., Klein, G. and Ernberg, I. (1971 a). *Virology* **45**, 10-21.
4. Gergely, L., Klein, G. and Ernberg, I., (1971 b). *Virology* **45**, 22-29.
5. Gergely, L., Klein, G. and Ernberg, I., (1971 c). *Int. Jl Cancer*, **7**, 293-302.
6. Hausen, H. zur, Schulte-Holthausen, H., Klein, G., Henle, W., Henle, G., Clifford, P. and Santesson, L. (1970). *Nature, Lond.* **228**, 1056-1058.
7. Henle, G. and Henle, W. (1966). *J. Bact.* **91**, 1248-1256.
8. Henle, W. and Henle, G. (1968). *J. Virol.* **2**, 182-191.
9. Henle, W., Henle, G., Zajac, B. A., Pearson, G., Waubke, R. and Scriba, M. (1970). *Science, N.Y.* **169**, 188-190.
10. Henle, G., Henle, W. and Klein, G., (1971). *Int. J. Cancer.* **8**, 272-282.
11. Klein, G. (1966). *A. Rev. Microbiol.* **20**, 223-252.
12. Klein, G., Gergely, L. and Goldstein, G. (1971). *Clin. expl. Immunol.* **8**, 593-602.
13. Klein, G. (1971). *Adv. Immunol.*, (in press).

14. Niderman, J. C., Evans, A. S., Subrahmanyan, L. and McCollum, R. W. (1970) *New Engl. J. Med.* **282**, 361-365.
15. Old, L. J., Boyse, E. A., Oettgen, H. F., de Harven, E., Geering, G., Williamson, B. and Clifford, P. (1966). *Proc. natn. Acad. Sci. U.S.A.* **56**, 1699-1704.
16. Pope, J. H., Scott, W., Reedman, B. M. and Walters, M. K. (1971). "Proceedings of the 1st Int. Symposium of the Princess Takamatsu Cancer Res. Fund" (K. Nishioka, ed.). Tokyo, (in press).
17. Silvestre, D., Kourilsky, F. M., Klein, G., Yata, Y., Neauport-Sautes, C. and Levy, J. P. (1971). *Int. Jl Cancer* **8**, 222-233.
18. Svedmyr, A., Demissie, A., Klein, G. and Clifford, P. (1970). *J. nat Cancer Inst.* **44**, 595-610.
19. Walters, M. K. and Pope, J. H. (1971). *Int. Jl Cancer* (in press).
20. Yata, J., Klein, G. (1969). *Int. Jl Cancer* **4**, 767.
21. Zajac, B. A. and Kohn, G. (1970). *J. nat Cancer Inst.* **45**, 339-406.

# Expression of the DNA-Tumour Virus Genome: Relationship to DNA Replication

G. SAUER, C. COLLINS, and H. FISCHER

*Institut für Virusforschung,
Deutsches Krebsforschungszentrum,
Heidelberg, Germany*

The DNA-tumour viruses interact with the host cells in two entirely different ways: either the cell is productively infected and all functions of the viral genome are expressed, resulting in the synthesis of viral progeny and lysis of the cell; or, during the transforming interaction of the virus with the cell, expression of the viral genome is curtailed in that only the so-called "early" viral functions are expressed, while "late" functions, such as synthesis of viral capsid proteins and virus maturation, are missing.

These differences in the expression of the viral genome led us to search for differences in transcriptional control in the two cell systems. Specifically, we have investigated the interrelationship between DNA synthesis and viral messenger RNA (mRNA) transcription. It appears that in a productive cycle of infection the early-late switch in mRNA transcription depends on the replication of the viral DNA [4, 9, 14]. While early functions can be transcribed from the incoming parental DNA, transcription of late viral mRNA requires DNA synthesis. In contrast, the transcription of late mRNA sequences in transformed cells appears to be independent of DNA replication [14].

We have also investigated this early-late switch in the expression of the viral genome in terms of the time-course of gene-product synthesis. By means of kinetic analysis and the use of inhibitors of DNA synthesis, we have been able to distinguish between early and late proteins in the infected cell. The appearance of late gene products again depends on the replication of the viral DNA.

The relationship between the expression of the viral genome and DNA synthesis prompted us to examine closely the possible dependence between DNA replication and the association of the viral DNA with various cell components during the initial stages of transformation.

In the studies to be described we have employed arabinofuranosylcytosine (ara-C) as an inhibitor of DNA synthesis. The SV40 monkey-cell system was used for productive infection and the SV40-3T3-cell system, which is nonpermissive for virus multiplication, was employed for the study of the initial interactions during transformation.

## TRANSCRIPTION AND PROTEIN SYNTHESIS DURING PRODUCTIVE INFECTION

To test whether the entire SV40 genome is transcribed at once during productive infection, or whether transcription occurs sequentially, the following hybridization competition experiment was performed (Fig. 1). Late $^3$H-labelled SV40 specific RNA was extracted from CV-1 cells at the end of the infectious cycle 50 hr post infection. A saturating amount of $^3$H-late RNA was added to the

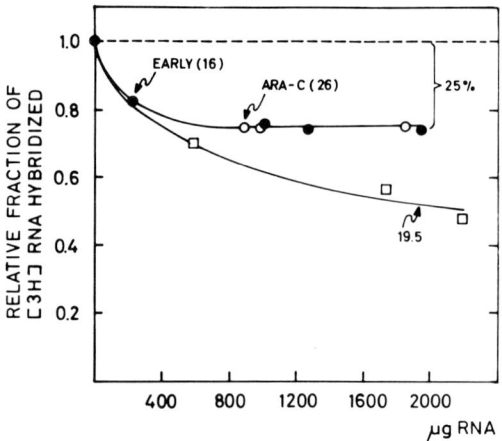

Figure 1. Requirement of DNA replication for transcription of late SV40 mRNA in productively infected cells. Hybridization competition was carried out between $^3$H-late SV40 mRNA extracted 50 hr post infection from productively infected CV-1 cells and unlabelled early SV40 mRNS extracted 16 hr post infection, SV40 mRNA obtained after treatment of infected cells for 26 hr with 12.5 μg/ml ara-C, and SV40 mRNA extracted at 19.5 hr post infection. The reaction mixture contained 0.05 μg of SV40-DNA, a saturating amount of 270 μg of $^3$H-late SV40 mRNA (specific activity 16,500 cpm/μg) and increasing amounts of the unlabelled RNA preparations (see ref. 14).

reaction mixture, which contained immobilized denatured SV40 DNA, and increasing amounts of early SV40-specific RNA extracted 16 hr post infection, i.e. prior to the onset of viral DNA replication [10].

The unlabelled early SV40 mRNA competes to the extent of 25% with the $^3$H-late SV40 mRNA. By contrast, after 19.5 hr post infection new SV40 mRNA sequences are synthesized which compete to a greater extent with the $^3$H-late SV40 mRNA. Since there are only a few copies of the newly transcribed

SV40 mRNA species present, the slope of the curve declines slowly. Addition of ara-C to the infected cultures throughout the period of infection prevents the formation of these new late mRNA sequences, as the RNA extracted after 26 hr from ara-C treated cultures competes with late $^3$H-SV40 mRNA to the same extent as RNA extracted at 16 hr post infection.

Several conclusions can be drawn from these data. First, transcription of the SV40 genome is sequential. At the early stages of infection only early SV40 mRNA sequences, which share 25% homology with late SV40 mRNA, are synthesized. Later, after the beginning of viral DNA replication, which occurs by 17 hr, both early and new later SV40 mRNA species are transcribed [14]. This agrees with observations described by Aloni et al. [1] and Oda and Dulbecco [13]. Furthermore, prevention of DNA synthesis by ara-C also inhibits completely the synthesis of late SV40 mRNA in productively infected cells.

This effect of ara-C on late mRNA transcription led us to study the appearance of late SV40 products such as viral capsid proteins in productively infected cells. The proteins of SV40 infected CV-1 cells were labelled for 30 min with $^3$H-amino-acids late in the infectious cycle, when the bulk of the virion proteins are synthesized. Separation of the proteins was done by SDS gel electrophoresis. $^{14}$C-labelled purified SV40 capsid proteins were added to the cellular proteins prior to electrophoresis in order to compare the electrophoretic mobility of the various peaks with the known SV40 marker proteins designated VP1 to VP6, which were also described by Girard et al. [7]. As shown in Fig. 2a, there appear late in the infectious cycle in the electrophoretic pattern four peaks from the infected cells which assume identical positions with the $^{14}$C-labelled SV40 capsid proteins. Treatment of the infected cells with ara-C completely prevents synthesis of these proteins (Fig. 2b).

Due to the time of appearance of these four proteins, the fact that their synthesis is inhibited by ara-C, and their co-migration with marker proteins from purified virus, we have classified these proteins as viral capsid proteins.

In addition to these four late protein peaks in the electrophoretic pattern, there are two virus-induced protein peaks which appear much earlier in the infectious cycle and which are insensitive to treatment by ara-C [5]. These early peaks, therefore, can be tentatively classified as non-capsid viral proteins.

It should be emphasized that the four major virus protein species described above account for approximately 90% of the coding capacity of the small SV40 genome. When one considers the number of virus-specific functions already established for SV40, e.g., initiation and maintenance of transformation and induction of T-antigen synthesis [3], then one is indeed faced with a coding dilemma in trying to account for the number of functions attributed to the SV40 genome. Among other possible explanations, it is tempting to speculate that the virus capsid proteins are capable of non-structural functions, or that they might arise as different cleavage products from one larger protein.

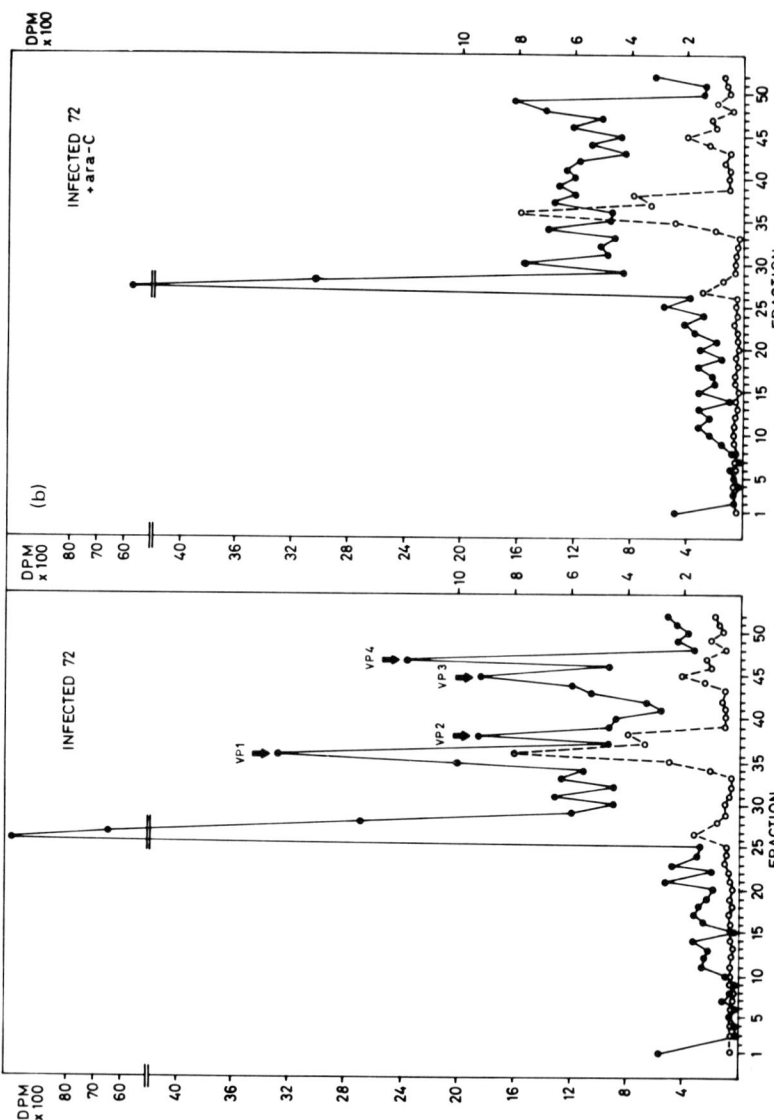

**Figure 2.** Polyacrylamide-gel-electrophoresis of solubilized proteins from SV40 infected CV-1 cells showing the protein synthetic activity at 72 hr after infection. Proteins of infected cells which had been labelled with $^3$H-amino-acids (●—●) for 30 min before harvesting were coelectrophoresed with $^{14}$C-labelled (○----○) purified SV40-capsid-proteins. Treatment with ara-C was carried out for 72 hr.

## TRANSCRIPTION IN TRANSFORMED CELLS

Cells transformed by SV40 are characterized as a rule by the presence of tumour-antigen, transplantation-antigen, SV40 mRNA [1, 13, 16], absence of viral capsid protein, and lack of production of infectious SV40. Some of these transformed cells contain early SV40 mRNA and SV40 mRNA which successfully competes to varying degrees with late SV40 mRNA from productively infected cells [12, 16]. There are, for example, SV40 transformed cell lines such as GMK-EVa cells [11] which, as shown by competition experiments, contain early SV40 mRNA and 10% of the late SV40 mRNA sequences

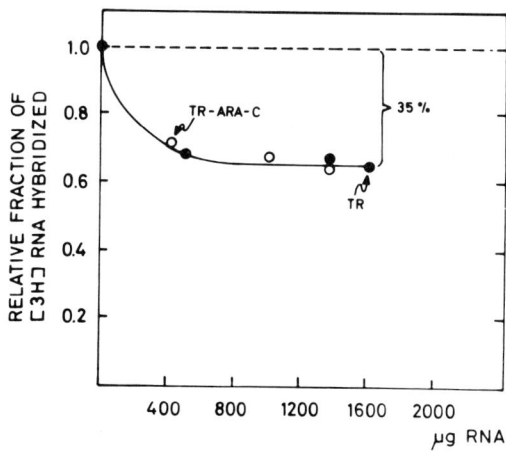

**Figure 3.** Synthesis of SV40 mRNA in SV40-transformed cells. Hybridization competition was carried out between $^3$H-late SV40 mRNA and SV40 mRNA extracted from SV40-transformed GMK-EVa cells either untreated or treated for 26 hr with 12.5 μg/ml of ara-C. The reaction mixture contained 0.05 μg/ml of SV40-DNA, $^3$H-late SV40 mRNA (see Fig. 1) and increasing amounts of unlabelled GMK-EVa RNA.

synthesized in productively infected cells. Despite the production of these late SV40 mRNA sequences, no late SV40 functions are detectable in these transformed cells. Such cell lines can be employed to investigate whether the mechanisms controlling the regulation of transcription of the SV40 genome are the same in productively infected cells and in transformed cells. If transcription of the late SV40 mRNA sequences in transformed cells proceeds even under conditions of complete inhibition of DNA synthesis, this would indicate essentially different control mechanisms of transcription between productively infected and transformed cells. It may be seen from Fig. 3 that addition of ara-C for 26 hr to GMK-EVa cells does not influence the synthesis of SV40 mRNA in these cells. In both the untreated and treated cells the homology between the SV40 mRNA and $^3$H-late SV40 mRNA proved to be 35% (25% representing early sequences and 10% late sequences). As previously pointed out, addition of ara-C to productively infected cells completely prevents late mRNA synthesis. Thus, the regulation of transcription of the SV40 genome appears to be different in productively infected and transformed cells, since late SV40 mRNA sequences can still be transcribed in transformed cells even under conditions of inhibited DNA synthesis [14].

It is possible that the recognition of late sections of the viral genome in productively infected cells by the DNA-dependent RNA-polymerase may be achieved by the availability of replicative forms of the viral DNA. Such molecules may reveal new initiation signals for the transcribing enzyme. These molecules are missing in transformed cells, which do not contain any infectious forms of the SV40 DNA [15]. The conversion in transformed cells of the infectious circular forms of the viral superhelical DNA to an open non-infectious molecule, which may be a prerequisite for integration into the cellular DNA, may also be one of the most likely explanations for the observed differences in transcriptional control. The conversion of the circular DNA into a linear integrated molecule might reveal new initiation signals for the DNA-dependent RNA-polymerase, which are missing in productively infected cells. These newly exposed signals may then allow the recognition of some of the late SV40 genes in the absence of DNA replication.

## DNA SYNTHESIS AND TRANSFORMATION

The exact nature of the events leading to structural changes in the infecting viral DNA, and possibly in the DNA of the host cell during the course of transformation, and the relationship of these conformational alterations to the integration process, are not clearly understood at present. It is known, however, that cellular DNA synthesis and mitotic divisions are required during the initial virus-cell interactions to ensure transformation of the infected cells. It has been shown that at least one round of cell division is necessary to fix the transformed state [17] and that the fraction of 3T3 cells induced by SV40 infection to replicate DNA and divide corresponds to those cells which become eventually transformed [6]. It may be speculated that cellular DNA replication might be a necessary step to render sites available within the cellular DNA helix for the integration of the infecting viral DNA.

We started, therefore, an investigation of the relationship between DNA synthesis and association of the donor SV40 DNA with various cellular components. For this purpose non-permissive BHK-21 and 3T3 cells have been infected with purified superhelical double-stranded SV40 DNA (form I). The use of non-permissive cells allows transformation rather than a lytic interaction, while infection with purified DNA instead of virions avoids complications due to uncoating events and allows a direct kinetic analysis of the fate of the viral DNA.

The kinetics of adsorption of $^{14}$C-SV40 DNA-I is shown in Fig. 4. Exposure of the cells for up to 40 min to the viral DNA leads to the binding of an increasing amount of the latter. After reaching this maximum level the curve seems to decline slightly. To avoid further adsorption of viral DNA after the period of infection had been terminated, and to ensure synchrony of the events after infection, the cultures were treated with DNase in order to digest the viral DNA which was not permanently adsorbed and therefore protected from the enzyme.

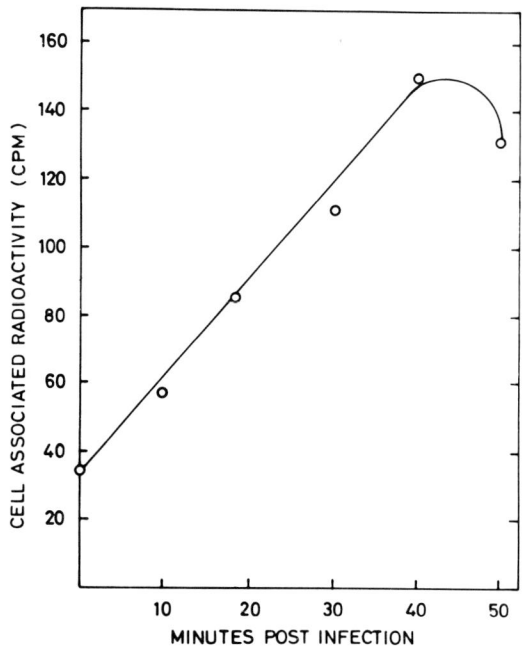

**Figure 4.** Confluent monolayers of BHK-21 in 6 cm petri dishes were infected with $^{14}$C-SV40-DNA-I (4 μg DNA, specific activity 2870 cpm/μg, 150 μg DEAE-Dextran in 0.2 ml Hanks solution). The adsorption was carried out at 37°C. At times indicated the cultures were treated with DNase (80 μg/plate), washed, harvested, and the cell-associated acid-insoluble radioactivity was determined.

To test the association of the viral DNA with various cellular components and the relationship of this association to DNA replication, the following general experimental technique was employed: First, confluent 3T3 cells were infected for 30 min with $^{14}$C-SV40 DNA-I, then, after DNase treatment, one-half of the infected cultures were released by trypsinization from contact-inhibition and reseeded at one-half of the original density. This resulted in the initiation of cellular DNA synthesis and mitotic division. The other cultures were treated with ara-C to inhibit DNA replication and, at various time intervals, the amount of the radioactively labelled SV40 DNA was determined in either the total cell fraction or in various cell components.

In order to prove that ara-C cultures are indeed blocked in cell division, the growth pattern of infected 3T3 cells, with and without ara-C treatment, was determined and the results are shown in Table 1. It may be seen that reseeded cultures undergo one cell division within two days after trypsinization, while ara-C treated cultures are completely prevented from growing.

Table 1. Growth pattern of 3T3 cells* after infection with SV40 DNA in presence or absence of ara-C.

| Hours after Trypsinization | Number of cells per plate ($\times 10^{-3}$) | |
|---|---|---|
| | with ara-C | without ara-C |
| 24 | 348 | 339 |
| 48 | 315 | 561 |
| 72 | 306 | 753 |
| 96 | 298 | 921 |

* Confluent 3T3 monolayers were infected as described in Fig. 5. All cultures were trypsinized 3 hr post infection and reseeded at one half of the original density. One group of cultures was then maintained in medium supplemented with 15 $\mu$g/ml of ara-C. At the times indicated, the mean number of cells was determined from duplicate cultures.

The amounts of total cell-associated radioactivity after various incubation times are shown in Fig. 5. The macromolecular acid-precipitable radioactivity which is associated with the whole cell declines rapidly within the first day post infection, but at later times the curve levels off and three days after infection the cells still harbour about 40% of the radioactivity initially present. The radioactivity of the cells which had undergone a cell division proved to be 10% greater than that remaining in resting cultures without DNA synthesis.

The investigation of the amount of incoming viral DNA associated with the nuclei of either dividing or resting (ara-C treated) cultures yielded a rather interesting result (Fig. 6). It can be seen that there is a considerable enhancement of binding of radioactivity to those nuclei that are allowed to undergo cellular DNA replication. This difference is most pronounced at two to three days post infection. At this time the dividing nuclei contained more than threefold the amount of viral DNA present in ara-C treated resting nuclei. Although the radioactivity again declines by the fourth day, it still remains considerably above the values measured for ara-C treated nuclei.

This enhanced binding of viral DNA to certain cellular components under conditions of cell growth was further confirmed by using another experimental technique. It has been shown by Hirt [8] that macromolecular cellular DNA can be effectively separated from viral DNA by precipitation and pelleting, after lysis of the cells with dodecylsulphate and treatment of the lysate with 1M NaCl. We therefore examined the association of the acid-precipitable viral DNA with the NaCl-dodecyl-SO$_4$-pellet from DNA synthesizing and resting cultures. Co-sedimentation of the radioactively labelled viral DNA with the cellular DNA under these conditions would suggest a close binding to the latter and a conversion of

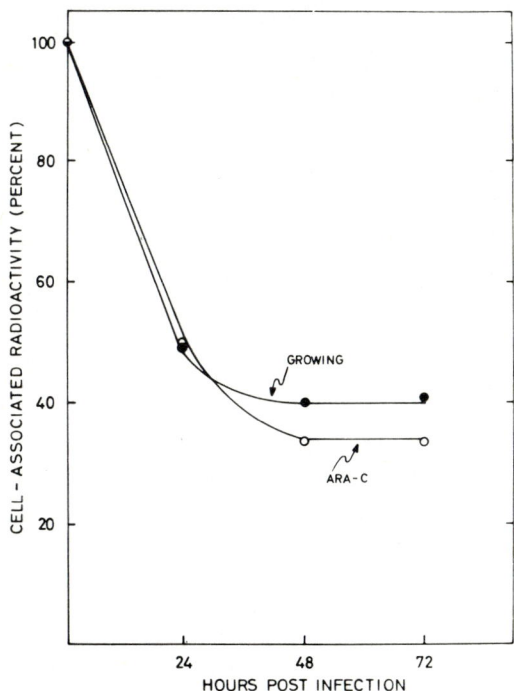

Figure 5. Confluent 3T3 monolayers were infected with $^{14}$C-SV40-DNA-I (31,300 cpm/4 µg DNA, 80 µg DEAE-Dextrab in 0.2 ml Hanks solution). Adsorption took place for 30 min at 37°C. The cultures were then DNase treated, washed and refed in medium supplemented with unlabelled thymidine. Non-growing cultures were maintained in medium containing 15 µg/ml ara-C; 3 hr post infection all cultures were trypsinized and reseeded at one half the original cell density. At times indicated the cells were harvested and the cell-associated acid-precipitable radioactivity determined.

the viral DNA from a circular closed conformation into another state not yet defined.

The data indeed show that the stimulation of cellular DNA replication in infected growing cultures renders considerably more viral DNA co-precipitable with the cellular DNA pellet than is found in ara-C inhibited cultures (Fig. 7).

These findings strongly suggest that cellular DNA synthesis facilitates a close association of the viral DNA with the cellular DNA. Whether this association reflects a stable integration event is now being studied under conditions which allow a clear determination of alkaline stable linkages between viral and cellular

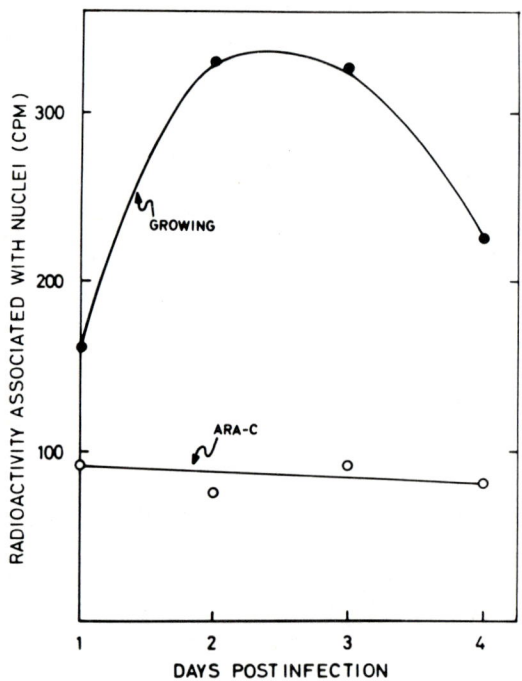

**Figure 6.** Confluent 3T3 monolayers were infected and further treated as described in Fig. 5. One half of the cultures were maintained after the infection procedure in medium containing 15 μg/ml ara-C. The remaining cultures were trypsinized and reseeded as described in Fig. 5. At times indicated the cells were harvested and nuclei isolated by treatment with NP 40 (see ref. 2). The acid-precipitable radioactivity associated with the nuclei was then determined. Since the non-growing cultures had been maintained without reseeding, the radioactivity indicated for the growing cultures represents the sum of the two reseeded plates and these values have been corrected for the loss of radioactivity due to cell loss upon replating.

DNA. If this dependence of integration on DNA synthesis were to be proven, and if it were also to be correlated with the yield of transforming cells, it would help to clarify the physiological conditions required for the integration of the tumour virus genes into the cellular genome, where they may then permanently express their transforming potentials.

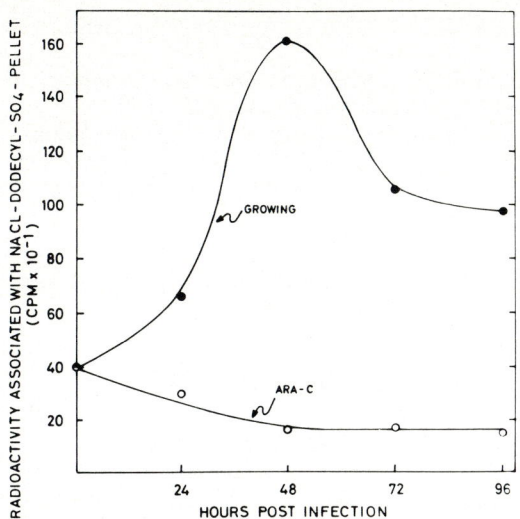

Figure 7. Confluent 3T3 monolayers were infected and further treated as described in Fig. 6. At times indicated the cells were washed and the NaCl-dodecyl-SO$_4$ pellets were obtained as described by Hirt (see ref. 8). The pellets were solubilized and the radioactivity determined. As pointed out in the legend to Fig. 6, the radioactivity indicated for the growing cultures represents the sum of the two reseeded plates.

## ACKNOWLEDGEMENTS

This work was supported by the Deutsche Forschungsgemeinschaft.

## REFERENCES

1. Aloni, Y., Winocour, E. and Sachs, L. (1968). *J. molec. Biol.* **31**, 415.
2. Barbanti-Brodano, G., Swetly, P. and Koprowski, H. (1970). *J. Virology* **6**, 78.
3. Black, P. H., Rowe, W. P., Turner, H. C. and Huebner, R. J. (1963). *Proc. natn. Acad. Sci., U.S.A.* **50**, 1148.
4. Butel, J. S. and Rapp, F. (1965). *Virology* **27**, 490.
5. Fischer, H. and Sauer, G. (1972). *J. Virol.* (in press).
6. Fox, T. O. and Levine, A. J. (1971). *J. Virology* **7**, 473.
7. Girard, M., Marty, L. and Suarez, F. (1970). *Biochem. biophys. Res. Commun.* **40**, 97.
8. Hirt, B. (1967). *J. molec. Biol.* **26**, 365.
9. Kit, S., Dubbs, D. R., Frearson, P. M. and Melnick, J. L. (1966). *Virology* **29**, 69.
10. Kit, S., Kurimura, T., Salvi, M. L. and Dubbs, D. R. (1968). *Proc. natn. Acad. Sci., U.S.A.* **60**, 1239.
11. Koprowski, H., Jensen, F. C. and Steplewski, Z. (1967). *Proc. natn. Acad. Sci., U.S.A.* **58**, 127.

12. Martin, M. A. and Axelrod, D. (1969). *Proc. natn. Acad. Sci., U.S.A.* **64**, 1203.
13. Oda, K. and Dulbecco, R. (1968). *Proc. natn. Acad. Sci., U.S.A.* **60**, 525.
14. Sauer, G. (1971). *Nature, Lond.* **231**, 136.
15. Sauer, G. and Hahn, E. C. (1970). *Z. Krebsforsch.* **74**, 40.
16. Sauer, G. and Kidwai, J. R. (1968). *Proc. natn. Acad. Sci., U.S.A.* **61**, 1256.
17. Todaro, G. J. and Green, M. (1966). *Proc. natn. Acad. Sci., U.S.A.* **55**, 302.

## DISCUSSION

**D. Luzzati:** What is the size of virus-specific mRNA late in infection and in transformed cells.

**G. Sauer:** The size of the SV40 specific mRNA extracted from productively infected cells varies over a wide range, comprising even giant messenger RNA molecules manyfold the size of the viral DNA.

# Protein Synthesis in Interferon-treated and Virus-infected Cells

IAN M. KERR, P. DOBOS,* E. M. MARTIN,† D. H. METZ, and M. ESTEBAN

*National Institute for Medical Research,*
*London, England*

## INTRODUCTION

In many virus-cell systems virus protein synthesis occurs while host protein synthesis is progressively inhibited (for a review see Martin and Kerr [43]). In contrast, in the interferon-treated infected cell, virus protein synthesis is inhibited while the majority of host protein synthesis is apparently unaffected (reviewed by Sonnabend and Friedman [57] and Vilcek [62]). The inhibition of host or viral protein synthesis could be at the level of the synthesis of the appropriate messenger RNA (mRNA) (transcription) or at the level of its translation into protein or both. In *E. coli* infected with T4 phage, for example, control of protein synthesis appears to be at both levels [1, 18]. Indeed, a new virus-specific polymerase, alterations in the host RNA polymerase or in the sigma factors associated with it (reviewed by Travers [61]), changes in host tRNAs [27, 59, 64] and in the enzymes which modify them [22, 63], the synthesis of new phage-coded tRNAs [11, 54, 60, 65], alterations in the aminoacyl tRNA synthetases [46] and in the factors controlling the initiation of translation of different mRNAs on the ribosome [14, 21, 34, 50, 51, 53, 58], have all been reported to occur in one or other of a variety of phage-infected systems. The interferon mediated inhibition of viral protein synthesis could in theory operate at any one of these levels and a number of others besides. For example, treatment of *E. coli* with colicin E3 results in a rapid inhibition of host protein synthesis. The colicin, a protein of molecular weight 60,000 daltons, does not enter the cell but its combination with specific sites on the cell membrane results in cleavage of the 16S ribosomal RNA, rendering the ribosome inactive in protein synthesis [3, 55]. Thus, combination of one or a few protein

* Present address: Department of Microbiology, Dalhousie University, Halifax, Nova Scotia, Canada.
† Present address: School of Biological Sciences, Flinders University, Bedford Park, South Australia.

molecules with specific sites at the cell membrane can produce dramatic structural and functional changes in remote components within the cell. Interferon treatment, in contrast to the action of colicin E3, would be expected to inhibit only virus protein synthesis. It is nevertheless tempting to postulate that a membrane-mediated modification might be involved in its mechanism of action.

Clearly, therefore, we have no shortage of possible mechanisms. In contrast we have little knowledge as to how many of them operate in animal cells and more particularly which, if any, are operative in interferon action. Different interferons are produced by different species and although cross reaction can occur, a given interferon is usually only active in cells of the same species as those in which it was produced—for example, human interferon in human cells and chick interferon in chick cells. No pure preparation of interferon is yet available, but the interferons appear from their physical properties and sensitivity to proteolytic enzymes to be protein in nature. They do not act directly upon the virus particle. On exposure of cells to interferon, resistance to infection with a wide spectrum of RNA and DNA viruses develops over a period of time. It is with the nature of the changes in the cell and how they confer this resistance that we are concerned in the study of interferon action. (A detailed account of the interferons is given by Vilcek [62] and by Finter [17]).

Studies employing inhibitors of RNA and protein synthesis led to the suggestion that interferon acts by inducing the cell to synthesize a new protein which inhibits the translation of viral but not of host cell mRNA (reviewed by Sonnabend and Friedman [57] and Vilcek [62]). While the synthesis of such a protein remains hypothetical, the data obtained by Joklik and Merigan [25] and by Levy and Carter [37] in their investigations of the fate of viral mRNA in the intact cell, are in accord with there being an inhibition of the translation of viral mRNA in the interferon-treated cell. The results of initial studies in the cell-free system also supported such a mechanism. Marcus and Salb [40, 41] concluded that ribosomes from interferon-treated cells would combine with, but not translate viral RNA, whereas Carter and Levy [8, 9] reported that such ribosomes neither bound viral RNA nor incorporated amino acids in response to it. In neither of the cell-free systems used, however, was the product synthesized in response to the viral RNA identified, and it now seems unlikely that the observations reported by Marcus and Salb are meaningful with respect to the mechanism of action of interferon (refs 31, 32, and P. I. Marcus, personal communication).

In the work to be described here, cell-free systems from mouse Krebs II ascites tumour and L cells and from chick embryo fibroblasts (CEFs) have been developed in which encephalomyocarditis (EMC) virus-specific polypeptides are synthesized in response to the viral RNA. The effect of interferon pretreatment of cells upon these systems has been examined. The results, while suggestive, do not provide definitive evidence for or against the hypothesis that virus protein synthesis is inhibited at the translational level in the interferon-treated cell. The data obtained from further studies concerning the processing of vaccinia virus mRNA in such cells, however, remain firmly in its favour.

## METHODS AND MATERIALS

The provision of cells, preparation of cell-free systems, treatment with partially-purified chick interferon, growth and purification of EMC virus, the isolation of EMC RNA and the characterization of the product synthesized in the cell-free system by electrophoretic and chromatographic analysis of tryptic digests and by electrophoresis on SDS-acrylamide gels were as described by Kerr and Martin [33], Kerr [32] and Dobos, et al. [13]. The procedures for the vaccinia virus (strain WR) experiments were similar to those described by Joklik and Merigan [25]. Caesium chloride gradients were as described by Baltimore and Huang [5]. Partially-purified mouse interferons were obtained from the Microbiological Research Establishment, Porton, England (see ref. 4) and from Dr. K. Paucker (see ref. 49); the specific activities were of the order of $10^6$ and $> 10^7$ units/mg protein respectively. The latter samples were the kind gift of Dr. R. M. Friedman.

## RESULTS AND DISCUSSION

The RNA genome of EMC virus (a picornavirus similar to polio, mengo and ME viruses) has a molecular weight of $2.7 \times 10^6$ daltons (see ref. 6) sufficient to code for a maximum of 270,000 molecular weight units of polypeptide. The virion capsid proteins appear to account for about one third of this coding capacity (refs 7, 13, 33, and Fig. 2), the remainder accounting for the non-capsid virus-specific polypeptides present only in the infected cell of which the viral RNA-dependent RNA polymerase(s) is probably an example. Stimulation of amino acid incorporation in cell-free systems by EMC RNA is well documented (Kerr et al. [29, 30] and Mathews and Korner [44]), but only recently has there been any progress in the analysis of the polypeptide product formed in response to the RNA. Smith et al. [56] compared tryptic digests of the products synthesized in such systems with those of virus coat protein and of virus protein synthesized in the infected cell. Although unable to detect the synthesis of virion coat protein peptides in response to the EMC RNA, these authors were able to show correspondence between peptides from the cell-free system and the infected cell. Moreover, the same polypeptide product in part appears to be synthesized in response to EMC RNA in cell-free systems from mouse and chick cells [33]. More recently, using higher resolution fingerprinting techniques, we have been able to show a much greater correspondence between the product synthesized in the cell-free system, the virion capsid proteins and the polypeptides present only in the infected cell. Typical analyses of tryptic digests of $^{35}$S-methionine labelled polypeptides from these three sources are shown in Figs. 1 and 2. This work has been presented in detail elsewhere [13]. Suffice it to say here that, from further analyses of peptides eluted from fingerprints of this

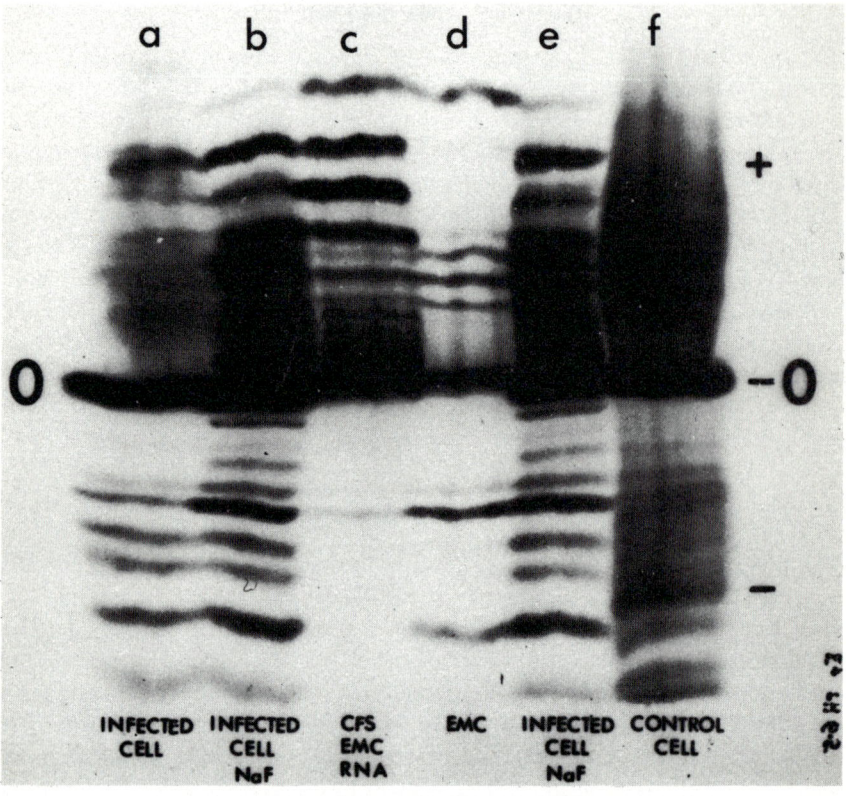

Figure 1. Analysis of virus-specific tryptic peptides by electrophoresis at pH 6.5 on thin layer silica gels. Samples of each digest were placed at 0 with the anode to the top of the figure. (a) protein from 3.5 hr EMC-infected Krebs cells labelled with $^{35}$S-methionine, without prior treatment with sodium fluoride. (b) and (e), protein from EMC-infected cells which were pre-treated with sodium fluoride (Methods). (c) product from the Krebs cell-free system supplemented with EMC RNA. (d) $^{35}$S-methionine-labelled EMC virus capsid protein. (f) protein from control cells labelled with $^{35}$S-methionine.

type (Fig. 2), fifty-two virus-specific methionine containing peptides were resolved in digests from infected cells. (This is about the number one would expect from complete translation of the viral genome into polypeptide containing 2% of methionine randomly distributed.) Of these about one third were also present in digests of the virion-capsid proteins (Fig. 2). The product

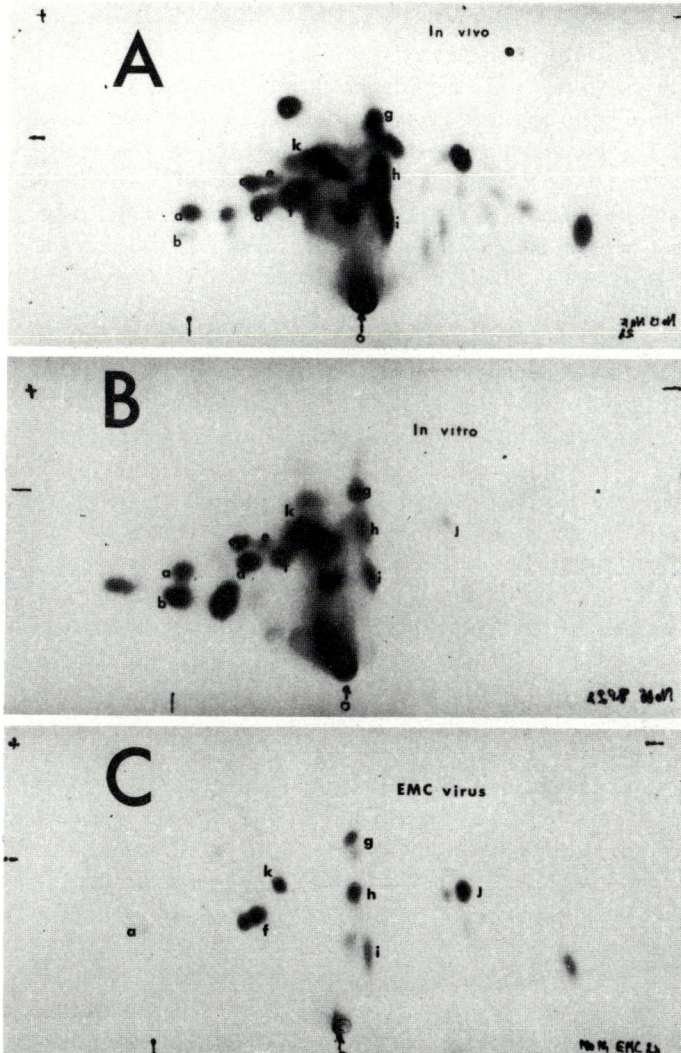

**Figure 2.** Fingerprint analysis of tryptic digests of $^{35}$S-methionine-labelled proteins from infected cells and cell-free systems: A. Digest of protein from EMC-infected Krebs cells labelled with $^{35}$S-methionine 3.5 hr post-infection after treatment with sodium fluoride. B. Digest from a Krebs cell-free system incubated with EMC RNA. Pre-incubated ribosomes were used and amino acid incorporation in the absence of added EMC RNA was into a large number of different peptides, none of which was sufficiently radioactive to be detected in the type of fingerprint shown here. C. Digest of purified EMC virus. The digests were placed at the bottom centre of the silica-gel sheets (arrow) and electrophoresed at pH 6.5, the anode is to the left. Chromatography was towards the top of the sheet. Autoradiography was for from 10 to 30 days. The line at the bottom left of each fingerprint indicates the distance moved by a phenol red marker during electrophoresis.

formed in response to EMC RNA in cell-free systems from Krebs mouse ascites tumour cells yielded 26 to 29 such peptides, the vast majority of which behaved identically to virus-specific peptides in digests from infected cells, while just under half of them appeared to be identical to peptides from the virion capsid proteins. The product formed in response to EMC RNA in cell-free systems prepared from L cells was similar, while Dobos *et al.* detected six additional EMC-specific peptides in mixed Krebs L cell systems [13]. It can be concluded that about 50% of the EMC RNA genome is being translated in these cell-free systems to yield products containing both virion capsid and non-capsid polypeptides.

It is of interest to ask why more of the genome is not translated. In the intact infected cell translation of the EMC genome appears to be similar to that of poliovirus [23, 24, 52] and several other picornaviruses [19]. The RNA is first translated to yield a high-molecular-weight precursor polypeptide(s) which is subsequently cleaved to yield the virion capsid and non-capsid proteins (ref. 7, and Dobos and Martin, Manuscript in preparation). Although initiation at secondary sites cannot yet be ruled out, these results are in accord with the initiation of virus protein synthesis at a unique site near the 5' end of the RNA from which the entire translatable section of the genome is then read. On this basis translation of up to 50% of the genome in the cell-free system would be expected to yield heterogeneous polypeptide(s) of up to 135,000 molecular weight. The result obtained by electrophoresis in SDS-acrylamide gels of the product synthesized in the cell-free system in response to EMC RNA is shown in Fig. 3. The product increases in size with time of incubation in the cell-free system up to a maximum of 85,000 to 100,000 daltons but, rather surprisingly, appears to form a number of discrete polypeptides the total molecular weight of which far exceeds the 270,000 molecular weight units for which the virus RNA carries information (Fig. 3). There must therefore be considerable overlap in the polypeptide sequences present. An analysis of tryptic digests of the polypeptide products formed in response to EMC RNA with time is shown in Fig. 4. The number of virus-specific peptides detectable increases with the time of incubation in the cell-free system. For example, the peptides in area *a1* are only detectable after 6 to 10 min and those in area *a2* after 11 to 40 min at 37°C. On the basis of these results sequential translation of the RNA genome could be occurring in these cell-free systems, the discrete products arising by specific cleavage of the RNA or of the nascent protein, or by intermittent termination at preferred sites on the RNA. Failure to translate the whole genome may reflect nucleolytic degradation of the RNA or simply failure to keep these systems functioning for more prolonged periods.

Accordingly, it can be concluded that a substantial portion of the EMC RNA genome is translated in these cell-free systems and that the results are in accord with sequential translation of the EMC RNA from a unique or very limited number of sites. That initiation of virus protein synthesis occurs exclusively at one site in the infected cell and, if so, that this is the only site used in the EMC RNA-stimulated cell-free system are points which still have to be firmly established, although recent work by Dr Alan Smith (personal communication) would suggest that this may well be the case.

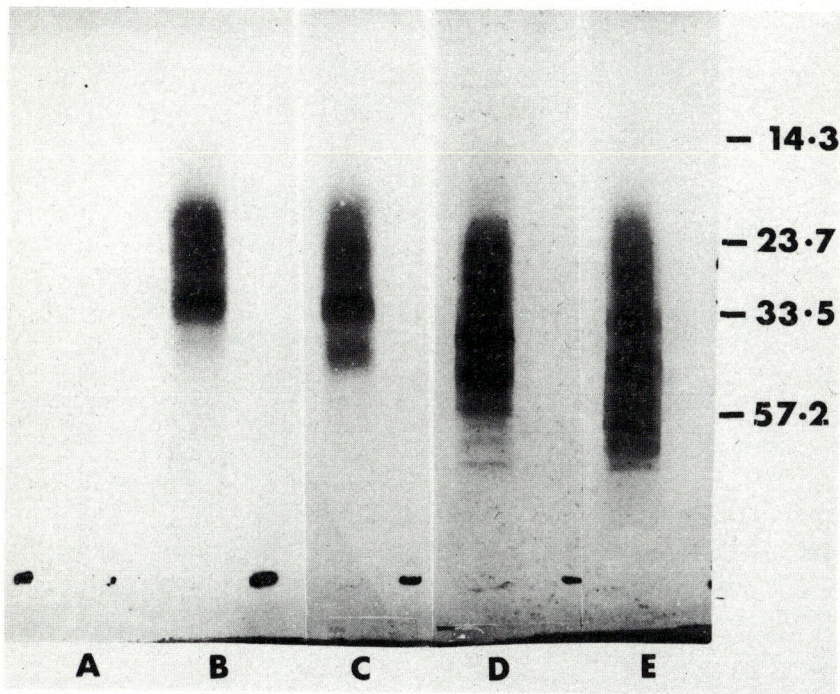

Figure 3. Analysis of the product synthesized in the cell-free system with time in the presence and absence of EMC RNA. After incubation of cell sap and preincubated ribosomes from Krebs cells with $^{35}$S-methionine, the protein from the cell-free system was analysed by electrophoresis on sodium dodecyl sulphate-acrylamide gels as described by Kerr and Martin (see ref. 33). A. Cell-free system incubated for 40 min at 37°C in the absence of EMC RNA. B, C, D, E. Cell-free systems incubated at 37°C in the presence of EMC RNA for 5, 10, 20 and 40 min respectively. The figures on the right indicate the molecular weights in thousands of marker proteins run in parallel gels.

What then is the effect of interferon pretreatment of cells on cell-free systems of this type? Unfortunately, in our hands the Krebs ascites tumour cell, with which much of the above work was done, responds very poorly to interferon and interferon inducers. Accordingly, we initially investigated the response to EMC RNA of cell-free systems from interferon-sensitive chick embryo fibroblasts (CEFs) and of mixed systems using components from both chick and mouse cells. We are heavily indebted to Dr Karl Fantes for provision of the partially-purified chick interferon [16] used throughout. We have previously

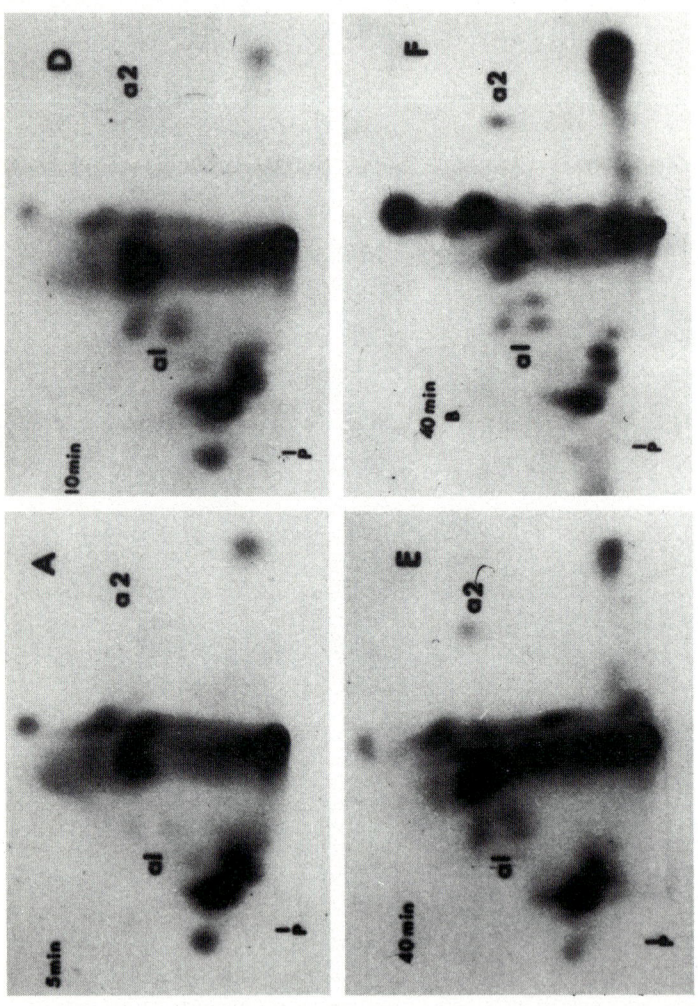

Figure 4. Fingerprint analysis of tryptic digests of the product synthesized in the cell-free system at different times of incubation. After incubation of cell sap and of preincubated ribosomes from Krebs cells with $^{35}$S-methionine and EMC RNA, the protein product was analysed by tryptic digestion and electrophoretic and chromatographic separation of the tryptic peptides as described in Fig. 2. Incubation in the cell-free system was for 5 min, A; 10 min, D; 40 min, E; and for 40 min in the presence of 1 mg/ml of bentonite, F.

shown that the same product, in part, is synthesized in response to EMC RNA in cell-free systems from CEFs and mouse cells, the difference being that in the CEF system less of the RNA appears to be translated [33]. In mixed systems containing components from mouse and chick cells, however, the product appears similar to that synthesized in the mouse systems.

A typical response to EMC RNA of ribosomes from control CEFs and CEFs which had been exposed to two different concentrations of interferon is shown in Fig. 5. The ribosomes used in these studies were a fraction including ribosome

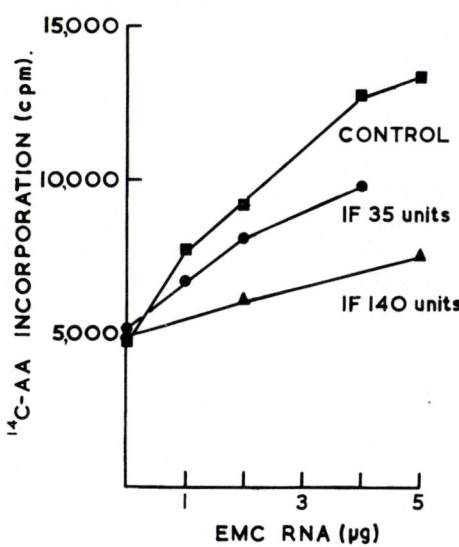

**Figure 5.** Response to encephalomyocarditis (EMC) RNA of ribosomes from interferon-treated and control cells. Stimulation of amino acid incorporation by EMC RNA with untreated "74S" ribosomes from control chick embryo fibroblasts (CEF, ■) and CEF which had been exposed to 35 (●) and 140 (▲) units of interferon per ml, was assayed in the presence of Krebs cell sap and a mixture of $^{14}$C-L-amino acids. Incubations were for 40 min at 37°C as described by Kerr and Martin (see ref. 33).

monomers, subunits and a few residual dimers isolated from the total microsome fraction by separation on sucrose density gradients without treatment with detergent. They were assayed in the presence of Krebs cell sap. The ribosomes from the interferon-treated cells responded less well to the EMC RNA than did the corresponding controls. The reduction in the extent of incorporation in response to EMC RNA observed with these ribosomes (Fig. 5) was paralleled by a similar reduction in rate. On the basis of fingerprint analysis of tryptic peptides, the product synthesized in response to the EMC RNA, however, appeared to be the same with both types of ribosome [32]. The decreased response to EMC RNA with material from interferon-treated cells must, therefore, reflect a reduction not in what is translated but in the rate or frequency of translation. The results with the cell sap fractions from interferon-treated cells were variable. Cell sap factors or factors arbitrarily distributed between the cell sap and ribosome fractions in different preparations are, however, specifically required for the translation of EMC RNA in these systems. As has previously been discussed [32] it would be premature, therefore, on the basis of these data to suggest that the effect of interferon pretreatment is upon the ribosome *per se*. With cell-free systems derived entirely from chick cells (as distinct from the mixed chick-mouse systems), those from interferon-treated cells were normally less active than corresponding controls. The restricted translation of EMC RNA observed in these latter systems however, although intriguing *per se*, limits their usefulness here and the use of mixed chick-mouse systems is not ideal [32]. The difference in response to EMC RNA observed with ribosomes from interferon-treated and control chick cells was clear and reproducible and the results of a number of control experiments were in accord with the effect being interferon mediated. Nevertheless, an effect of a contaminant in the partially-purified interferon cannot be entirely excluded nor, even if interferon mediated, do we have any means of knowing whether the effect observed is primary to interferon action or a mere secondary consequence of it. Unfortunately, a test of the activity of the ribosomes with an added non-viral mRNA other than synthetic polynucleotides has not been possible; for example, neither mouse nor rabbit haemoglobin mRNA or messenger ribonucleoprotein (mRNP) (the kind gifts of Dr. R. Williamson) appear to be active under the conditions tested so far in our chick systems. In view of these uncertainties, confirmation in other systems of the effect observed with the chick system would clearly be desirable. Accordingly, with the recent interest and advances in the purification of mouse interferon, we have turned our attention to cell-free systems from interferon-sensitive mouse L cells. These cell-free systems respond to EMC RNA much as do those from Krebs cells both quantitatively and with respect to the nature of the polypeptide product synthesized in response to the EMC RNA as determined by fingerprint analysis of tryptic peptides [13].

Two partially-purified preparations of mouse interferon of specific activity $10^6$ and $10^7$ units/mg of protein (Materials) have been used. No striking difference has been observed in the response to EMC RNA of cell-free systems from L cells exposed to 25 units/ml of either of these interferons, as compared with corresponding controls. Both ribosome and cell sap fractions from such interferon-treated cells are capable of supporting the translation of EMC RNA. Small differences in activity of the type one might expect from the results with the chick cell systems cannot, however, be excluded. For example, in experiments designed to follow the fate of EMC RNA in these cell-free systems preliminary results suggest that there may be some difference in the specific binding of EMC RNA to proteins of the cell sap fractions from interferon-treated and control cells (Martin, unpublished results). Also the concentration of interferon used in these studies has so far been lower than that routinely used with the chick cells.* Nevertheless, it seems probable that any difference that we are able to detect in these cell-free systems is likely to be small. This is in marked contrast to the results of Carter and Levy [9], who, also working with L cells and the closely related mengo virus RNA, demonstrated that ribosomes from interferon-treated cells showed a reduced ability to bind viral RNA and a negligible incorporation of amino acids in response to it. The reason for these conflicting results is not clear. It could simply lie in subtle distinctions in the mode of preparation of the cell-free systems. Clearly, what is required is a definition of the number and nature of the factors controlling the translation of EMC RNA in these systems and which are rate-limiting under the conditions employed. Quite apart from the problem posed by the possible distinction in roles of membrane-bound and free ribosomes, it is possible that a factor limiting the rate of translation of viral RNA in the interferon-treated intact cell may no longer be rate limiting in a cell-free system, during the preparation of which a more labile component may have been preferentially destroyed. Far, therefore, from being able, as we had originally hoped, to use interferon as a tool to investigate the factors controlling translation in the animal cell, we may well have to understand these latter thoroughly before the interferon question will be resolved.

To date, therefore, these cell-free system studies have failed to provide a conclusive answer as to whether or not virus protein synthesis is inhibited at the translational level in the interferon-treated cell. On the basis of our results concerning the processing of vaccinia virus RNA in the intact cell, however, this hypothesis remains as attractive as ever.

Vaccinia virus is a large DNA virus which multiplies in the cytoplasm of the infected cell (see McAuslan [39] for a recent review). It consists essentially of a core containing the DNA genome and an RNA polymerase (transcriptase) surrounded by an outer envelope. On entering the cell the envelope is removed and the core with its polymerase is liberated into the cytoplasm ("first stage uncoating"). The enzyme is activated and a burst of viral mRNA synthesis occurs, reaching a peak (when a multiplicity of several hundred virus particles per cell is used) between 30 and 60 min post infection. Proteins made under the instruction of this early mRNA serve both to switch off its own synthesis and to liberate the DNA from the viral core ("second stage uncoating"). Following this, more viral mRNA is made which is believed to code for the viral DNA

* Only limited amounts of these highly-purified interferon preparations have been available to us and the use of crude material has, in our experience, resulted on occasion in gross effects on the intrinsic activity of the cell-free systems as well as in their response to polyuridylic acid and EMC RNA—effects unlikely to be related to interferon action.

polymerase. The viral genome is then replicated, the late viral mRNA is made and translated and the progeny virions assembled.

It is the synthesis and processing of the earliest viral mRNA in the cytoplasm of the infected cell that is especially suitable for the study of the mechanism of action of interferon since it can be followed by short pulses of radioactive label, adopting conditions under which rapidly labelled host mRNA remains confined to the nucleus. This synthesis of early viral mRNA is insensitive to moderate doses (1 $\mu$g/ml) of Actinomycin D (Metz, unpublished results). Presumably the drug does not penetrate the viral core. Thus, viral mRNA can be labelled for extended periods while the synthesis of cellular messenger and ribosomal RNA is completely suppressed. In addition, it is of particular interest with respect to the interferon problem that this early viral mRNA is transcribed by the pre-existing virion core polymerase, for its synthesis is thus independent of prior virus protein synthesis. Indeed, if protein synthesis is inhibited, for example by cycloheximide, second stage uncoating does not occur, the core remains intact and early viral mRNA synthesis continues for an extended period. Interferon pretreatment has a similar effect. Despite the fact that more viral mRNA is made in the interferon treated cell, however, a smaller proportion of it associates with ribosomes to form virus-specific polysomes than in corresponding controls (Fig. 6). The profile of the polysomes from untreated infected cells is unremarkable both with regard to the distribution of the host ribosomes (the A260 trace) and of the viral mRNA ($^3$H-RNA trace). In contrast, the profiles for the interferon-treated cells are characteristic of a situation in which protein synthesis is inhibited. The ribosomes run off the polysomes and accumulate as single ribosomes, re-initiation of the cycle and passage of the $^3$H-mRNA into the polysome region are inhibited (Fig. 6).

These results suggest that the processing rather than the synthesis of the viral mRNA is blocked in the interferon-treated cell and are essentially similar to those previously described by Joklik and Merigan [25]. Accordingly, we have examined in detail this processing of the viral mRNA from its synthesis to the point at which it becomes associated with at least one ribosome. Cytoplasmic structures containing viral mRNA were identified first by sedimentation in sucrose gradients (Fig. 7A) and subsequently individual fractions from these gradients were further analysed by density in caesium chloride gradients. An example of one such analysis is shown in Fig. 7B. Putting all of the data from these different gradients together, it can be concluded that, under the condition of extraction used, the viral mRNA does not exist free in the cytoplasm of the infected cell. It is present primarily in the form of a ribonucleoprotein (RNP) complex of density 1.40 g/ml, the majority of which sediments heterogeneously at about 30S. (Early viral mRNA when freed from protein sediments in a broad peak at 13S.) A second component containing viral mRNA sediments slightly faster than the peak of single ribosomes and has a density of 1.48 g/ml compared with 1.52 g/ml for the ribosomes *per se*. As more ribosomes become associated with the mRNP the density increases to 1.52 as expected. A further component of density 1.46 g/ml having an S value between 50 and 80S is likely to be the initiation complex formed by combination of the mRNP with the 40S ribosome subunit.

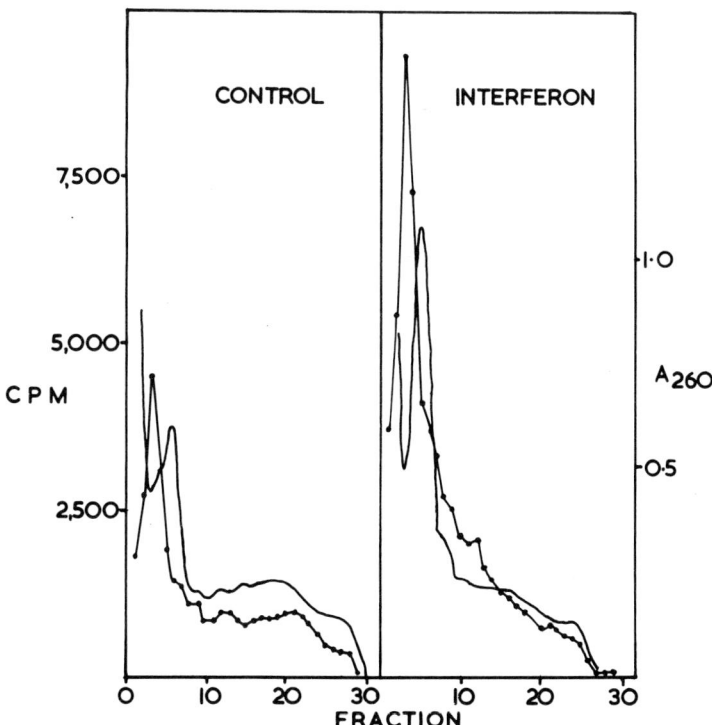

**Figure 6.** The formation of vaccinia virus specific polysomes in interferon-treated and control cells. 20 x $10^6$ L cells at a concentration of 1.4 x $10^6$ ml were treated with 50 units/ml of interferon (as assayed in the same cell-virus system) for 4 hr, diluted four-fold and incubated for a further 16 hr at 37°C. A batch of control cells was similarly treated without the addition of interferon. After infection with purified vaccinia virus at an added multiplicity of 500 particles per cell, the cells were pulse labelled with $^3$H-uridine (specific activity 30 Ci/mmole) from 30 to 50 min post-infection. Cytoplasmic extracts were centrifuged on 5 to 30% sucrose gradients at 130,000 x g for 45 min at 4°C (sedimentation was from left to right).
(———) optical density at 260 nm; (●) $^3$H-uridine.

Thus we believe that the early viral mRNA, after synthesis in the cores, associates with cellular protein to form a RNP complex: this, in the presence of appropriate factors and initiator tRNA, combines with the 40S ribosomal subunit, a 60S subunit being subsequently added to form the monosome. In the interferon-treated infected cell the mRNP is made but its association with

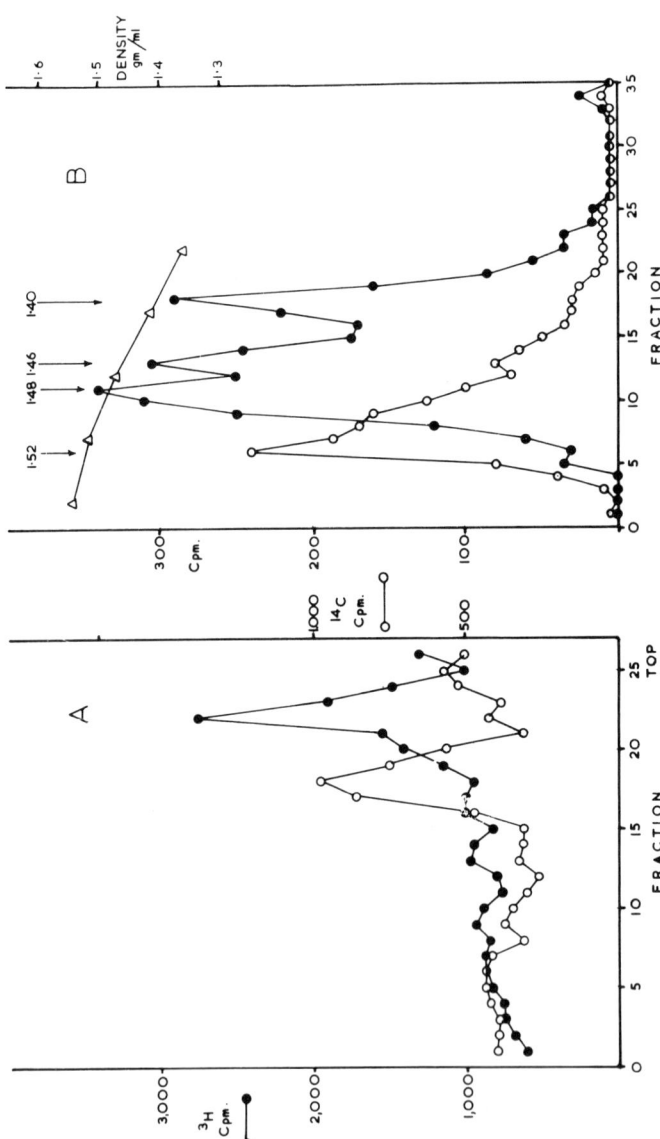

**Figure 7.** Analysis of the distribution of vaccinia virus early mRNA in the cytoplasm of the infected cell. A. $80 \times 10^6$ L cells were grown in suspension overnight in the presence of 5 μCi of $^{14}$C-uridine (specific activity 60 Ci/mole), infected with purified vaccinia virus at a multiplicity of 500 particles per cell, diluted to $2 \times 10^6$ cells/ml with Eagle's medium containing 5% calf serum and 1 μg/ml of Actinomycin D and pulse labelled from 20 to 40 min post-infection with 300 μCi of $^3$H-uridine (specific activity 30 Ci/mmole). A cytoplasmic extract was prepared and centrifuged on a 5 to 30% sucrose gradient at 90,000 x g for 2.5 hr at 4°C. (●) $^3$H-uridine; (○) $^{14}$C-uridine. B. Material from fraction 16 of the sucrose gradient shown in A was fixed with gluteraldehyde (see ref. 5) and centrifuged in a caesium chloride gradient at 100,000 x g for 17 hr at 4°C. In both A and B sedimentation was from right to left.

ribosomes to form polysomes is inhibited. We can ask, therefore, whether it is the addition of the small or of the large ribosomal subunit that is blocked. It should be possible to determine this from the relative amounts of mRNP, initiation complex and monosomes seen in the caesium chloride gradients of fractions from interferon-treated and control cells. This is under investigation. It is apparent, however, that the inhibition could be the result of a modification of one of the ribosomal subunits, of the initiation or as yet unrecognized factors, of the proteins of the mRNP or of the mRNA itself.

Whatever the explanation, it is clear that in the interferon-treated cell what appears to be vaccinia mRNA is synthesized but not translated. With the recent excitement concerning the possible "ticketing" of mRNA by the addition of polyadenylic acid [12, 15, 28, 36, 38] it is interesting to postulate an alteration at this level in the interferon-treated cell. As always, one would have to provide for selection between host and viral mRNAs, but there is no data which excludes the possibility that the processing of a fraction of host mRNA might not also be affected in the interferon-treated cell without the overall levels of RNA and protein synthesis being affected. Be this as it may, either a factor required for the recognition of vaccinia mRNA or for its alteration prior to translation is defective in the interferon-treated cell, or the RNA is not what it appears to be but nonsense produced by an altered viral polymerase. (Control experiments have ruled out significant contamination of this fraction by "leakage" of host RNA from the nucleus of the infected cell.) The putative mRNA made on vaccinia DNA even by an altered polymerase remains likely to hybridize with it and in the absence of a stringent test for messenger function its "meaningfulness" may prove difficult to establish. Mere stimulation of amino acid incorporation in the cell-free system would not be an adequate test: some analysis of the product would be essential and likely to prove complex. The importance of this question, however, has been emphasized not only by our own difficulties in obtaining definitive evidence in cell-free systems for the translation inhibition hypothesis, but also by two recent reports which, although inconclusive, reassert the possibility that in the interferon-treated cell transcription is affected. Firstly, Oxman and Levin [47] have shown that interferon pretreatment of monkey cells acutely infected with Simian Virus 40 (SV40) reduced the content of early virus-specific presumptive mRNA as well as the content of early viral protein (T-antigen). This unexpected finding suggests that either the action of interferon involves the transcription of early genes or, if indeed interferon acts at the level of translation, that the SV40 genome contains "proto-early" genes which must be transcribed and translated for transcription of the remaining early genes to occur [47]. Secondly a preliminary communication of Marcus et al. [42] suggests that the activity of the RNA polymerase (transcriptase) present in the vesicular stomatitis virus (VSV) particle is inhibited in the interferon-treated VSV-infected cell. A similar effect has been observed by

Huang and her colleagues (Huang, personal communication). With the relatively recent realization that, in phage-infected bacteria at least, host in addition to viral components may be used in the formation of a viral polymerase responsible for the production of viral mRNA [2, 26, 35, 61], it becomes easier to conceive of a unique interferon-mediated alteration in a host factor modifying the viral RNA polymerases (transcriptases) of different viruses. Similarly, the demonstration of the involvement in bacteria of cAMP and of ppGpp in the selection between the synthesis of ribosomal and mRNAs suggests how different polymerases all regulated by similar co-factors might be controlled [10, 45, 48]. It must be emphasized, however, that to date there is no evidence for the involvement of any host component in the activity of animal virus polymerases and one could equally plausibly invoke the operation of similar hypothetical mechanisms at the translational level. Certainly if there is an inhibition of the transcription of VSV RNA in the infected cell, this would be in direct contrast to the apparent situation with vaccinia virus and would, provided the virus RNA made in the interferon-treated vaccinia-infected cell is meaningful messenger, suggest the by no means impossible conclusion that alteration in a given factor(s) affects both transcription and translation of viral RNA in the interferon-treated cell. As Oxman and Levin [47] point out, however, there are a number of possible explanations for their results and it would be premature to reach any conclusion on the basis of the preliminary reports with VSV.

In summary, therefore, with respect to our own work, we have developed cell-free systems capable of the translation of a substantial portion of the EMC RNA genome. Initiation of virus protein synthesis in these systems appears to occur as in the infected cell, although this remains to be firmly established. Despite this these systems have failed to provide definitive evidence concerning the translation inhibition hypothesis of interferon action, while observations by other groups at the level of the intact cell have suggested a possible effect of interferon on transcription. Our own results in the intact cell, however, in extension of those of Joklik and Merigan [25] and of Horak *et al.* [20], remain in accord with there being an inhibition of the processing (translation in the widest sense of the word) of vaccinia virus mRNA in the interferon-treated cell.

Perhaps in our current state of knowledge, by analogy with the phage system, the most likely hypothesis remains that in the interferon-treated cell there is an alteration, quantitative or qualitative, in a factor of the F3 (B) type governing selection between different mRNAs [50, 51], which favours the translation of host but not viral RNA. It will be interesting, therefore, to see if EMC RNA is translated by cell-free systems from vaccinia-infected cells and, if so, whether interferon pretreatment affects this. One deficiency of the majority of the cell-free system studies carried out to date is that they have been carried out with material from interferon-treated rather than interferon-treated *infected* cells while it is at least possible that the full interferon-mediated response might not develop in the cell unless "triggered" by infection. Fundamentally, however, what is required is knowledge as to the factors controlling the activity of the viral RNA polymerases-transcriptases, whether viral RNA has to be modified in any way prior to translation and what factors are required for its translation. Cell-free systems of the type described here, and similar systems currently being developed elsewhere, should provide the means to ask the questions. As the answers become available and our understanding of the control of macromolecular synthesis in the infected cell gradually develops, the mechanism of action of interferon will no doubt slip quietly into place.

## REFERENCES

1. Adesnik, M. and Levinthal, C. (1970). *J. molec. Biol.* **48**, 187.
2. August, J. T. (1969). *Nature, Lond.* **222**, 121.
3. Bowman, C. M., Dahlberg, J. E., Ikemura, T., Konisky, J. and Nomura, M. (1971). *Proc. natn. Acad. Sci. U.S.A.* **68**, 964.
4. Bradish, C. J. and Allner, K. (1970). *In* "International Symposium on Standardisation on Interferon and Interferon Inducers, London 1969" Symposium Series Immunobiological Standardisation, (F. T. Perkins and R. H. Regamey, eds), pp. 35-36, Vol. 14, Karger, Basel and New York.
5. Baltimore, D. and Huang, A. S. (1968). *Science, N.Y.* **162**, 572.
6. Burness, A. T. H. (1970). *J. gen. Virol.* **6**, 373.
7. Butterworth, B. E. Stoltzfus, C. M. and Rueckert, R. R. (1971). *Fedn. Proc. Fedn. Am. Socs. exp. Biol.* **30**, 1216 Abs. Abstract No. 959.
8. Carter, W. A. and Levy, H. B. (1967). *Science, N.Y.* **155**, 1254.
9. Carter, W. A. and Levy, H. B. (1968). *Biochim. biophys. Acta* **155**, 437.
10. de Crombrugghe, B., Chen, B., Anderson, W., Nissley, P., Gottesman, M. and Pastan, I. (1971). *Nature New Biology Lond.* **231**, 139.
11. Daniel, V., Sarid, S. and Littauer, U. Z. (1968). *FEBS Letters* **2**, 39.
12. Darnell, J. E., Wall, R. and Tushinski, R. J. (1971). *Proc. natn. Acad. Sci. U.S.A.* **68**, 1321.
13. Dobos, P., Kerr, I. M., and Martin, E. M. (1971). *J. Virol.* **8**, 491.
14. Dube, S. K. and Rudland, P. S. (1970). *Nature, Lond.* **226**, 820.
15. Edmonds, M., Vaughan, M. and Nakazato, H. (1971). *Proc. natn. Acad. Sci. U.S.A.* **68**, 1336.
16. Fantes, K. H. (1967). *J. gen. Virol.* **1**, 257.
17. Finter, N. B. (ed.) (1966). "Interferons", North Holland Publishing Co., Amsterdam.
18. Friesen, J. D., Dale, B., and Bode, W. (1967). *J. molec. Biol.* **28**, 413.
19. Holland, J. J. and Kiehn, E. D. (1968). *Proc. natn. Acad. Sci. U.S.A.* **60**, 1015.
20. Horak, I., Hilfenhaus, J., Siegert, W., Jungwirth, C., Bodo, G. and Palese, P. (1970). *Z. Naturf.* **25b**, 1164.
21. Hsu, W.-T. and Weiss, S. B. (1969). *Proc. natn. Acad. Sci., U.S.A.* **64**, 345.
22. Hsu, W.-T., Foft, J. W. and Weiss, S. B. (1967). *Proc. natn. Acad. Sci. U.S.A.* **58**, 2028.
23. Jacobson, M. F. and Baltimore, D. (1968). *Proc. natn. Acad. Sci. U.S.A.* **61**, 77.

24. Jacobson, M. F., Asso, J. and Baltimore, D. (1970). *J. molec. Biol.* **49**, 657.
25. Joklik, W. K. and Merigan, T. C. (1966). *Proc. natn. Acad. Sci. U.S.A.* **56**, 558.
26. Kamen, R. (1970). *Nature, Lond.* **228**, 527.
27. Kan, J., Nirenberg, M. W. and Sueoka, N. (1970). *J. molec. Biol.* **52**, 179.
28. Kates, J. (1970). *Cold Spring Harb. Symp. quant. Biol.* **35**, 743.
29. Kerr, I. M., Martin, E. M., Hamilton, M. G. and Work, T. S. (1962). *Cold Spring Harb. Symp. quant. Biol.* **27**, 259.
30. Kerr, I. M., Cohen, N. and Work, T. S. (1966). *Biochem. J.* **98**, 826.
31. Kerr, I. M., Sonnabend, J. A. and Martin, E. M. (1970). *J. Virol.* **5**, 132.
32. Kerr, I. M. (1971). *J. Virol.* **7**, 448.
33. Kerr, I. M. and Martin. E. M. (1971). *J. Virol.* **7**, 438.
34. Klem, E. B., Hsu, W.-T. and Weiss, S. B. (1970). *Proc. natn. Acad. Sci. U.S.A.* **67**, 696.
35. Kondo, M., Gallerani, R. and Weissman, C. (1970). *Nature, Lond.* **228**, 525.
36. Lee, S. Y., Mendecki, J. and Brawerman, G. (1971). *Proc. natn. Acad. Sci. U.S.A.* **68**, 1331.
37. Levy, H. B. and Carter, W. A. (1968). *J. molec. Biol.* **31**, 561.
38. Lim, L. and Canellakis, E. S. (1970). *Nature, Lond.* **227**, 712.
39. McAuslan, B. R. (1969). *In* "The Biochemistry of Viruses" (H. B. Levy, ed.) pp. 361-408, Marcel Dekker, New York and London.
40. Marcus, P. I. and Salb, J. M. (1966). *Virology* **30**, 502.
41. Marcus, P. I. and Salb, J. M. (1967). *In* "The Interferons" (G. Rita, ed.), pp. 111-127. Academic Press Inc., New York and London.
42. Marcus, P. I., Engelhardt, D. L., Hunt, J. M. and Sekellick, M. J. (1971). "American Soc. Microbiology, Bacteriological Proceedings 1971, Abstracts 71st Annual Meeting, p. 213. Abstract Number V275.
43. Martin, E. M. and Kerr, I. M. (1968). *In* "Eighteenth Symposium The Society for General Microbiology" (L. V. Crawford and M. G. P. Stoker, eds), pp. 15-46. Cambridge University Press, Cambridge, England.
44. Mathews, M. B. and Korner, A. (1970). *Eur. J. Biochem.* **17**, 328.
45. Miller, Z., Varmus, H. E., Parks, J. S., Perlman, R. L. and Pastan, I. (1971). *J. biol. Chem.* **246**, 2898.
46. Neidhardt, F. C. and Earhart, C. F. (1966). *Cold Spring Harb. Symp. quant. Biol.* **31**, 557.
47. Oxman, M. N. and Levin, M. J. (1971). *Proc. natn. Acad. Sci. U.S.A.* **68**, 299.
48. Parks, J. S., Gottesman, M., Perlman, R. L. and Pastan, I. (1971). *J. biol. Chem.* **246**, 2419.
49. Paucker, K., Berman, B. J., Golgher, R. R. and Stancek, D. (1970). *J. Virol.* **5**, 145.
50. Pollack, Y., Groner, Y., Aviv (Greenshpan), H. and Revel, M. (1970). *FEBS Letters* **9**, 218.
51. Revel, M., Aviv (Greenshpan), H., Groner, Y. and Pollack, Y. (1970). *FEBS Letters* **9**, 213.
52. Roumiantzeff, M., Summers, D. F. and Maizel, J. V. Jr. (1971). *Virology* **44**, 249.
53. Schedl, P. D., Singer, R. E. and Conway, T. W. (1970). *Biochem. biophys. Res. Commun.* **38**, 631.
54. Scherberg, N. H. and Weiss, S. B. (1970). *Proc. natn. Acad. Sci. U.S.A.* **67**, 1164.

55. Senior, B. W. and Holland, I. B. (1971). *Proc. natn. Acad. Sci. U.S.A.* **68**, 959.
56. Smith, A E., Marcker, K. A. and Mathews, M. B. (1970). *Nature, Lond.* **225**, 184.
57. Sonnabend, J. A. and Friedman, R. M. (1966). *In* "Interferons" (N. B. Finter, ed.), pp. 202-267. North Holland Publishing Co. Amsterdam.
58. Steitz, J. A., Dube, S. K. and Rudland, P. S., (1970). *Nature, Lond.* **226**, 824.
59. Sueoka, N. and Kano-Sueoka, T. (1964). *Proc. natn. Acad. Sci. U.S.A.* **52**, 1535.
60. Tillack, T. W. and Smith, D. W. E. (1968). *Virology* **36**, 212.
61. Travers, A. (1970). *Nature New Biology Lond.* **229**, 69.
62. Vilcek, J. (1969). "Interferon" Virology Monograph 6. Springer-Verlag, Wien and New York.
63. Wainfan, E., Srinivasan, P. R. and Borek, E. (1965). *Biochemistry, N.Y.* **4**, 2845.
64. Waters, L. C. and Novelli, G. D. (1967). *Proc. natn. Acad. Sci. U.S.A.* **57**, 979.
65. Weiss, S. B., Hsu, W.-T., Foft, J. W. and Scherberg, N. H. (1968). *Proc. natn. Acad. Sci. U.S.A.* **61**, 114.
66. Bowman, C. M., Sidikaro, J. and Nomura, M. (1971). *Nature New Biology* **234**, 48, 133-137.

## DISCUSSION

**J. Werenne:** Have you attempted to study the binding of viral mRNA on ribosomes from interferon-treated cells and normal cells in the system of Heywood?

**I. M. Kerr:** We have studied binding of viral RNA to ribosomes under the conditions of Marcus and Salb and on cell-free systems, but not under those of Heywood. In our experience, specific binding at $0°C$ is masked by non-specific, and we have not observed any difference in binding with cell-free systems from interferon-treated and control cells. Incidentally, Dr. Marcus had concluded, from more recent data, that the binding studied by his group bears no relation to the activity of the RNA in protein synthesis.

**J. Drews:** Has anyone attempted to isolate an initiation complex in these cell-free systems comprising the initiation site of the viral RNA as well as the initiator tRNA and the ribosome?

**I. M. Kerr:** Dr. Alan Smith has isolated the initiation peptide in the cell-free system and in the intact infected cell, and has shown that initiation occurs in both cases with methionine, alanine, and (I think) threonine or proline. He is currently trying to isolate the initiation site on the viral RNA and will be discussing this here during the symposium on protein synthesis.

**J. Drews:** Can you say something about the relative efficiencies of the three cell lines, and the respective cell-free systems, in supporting the translation of EMC virus RNA?

**I. M. Kerr:** All that we can really say at the moment is that less of the EMC RNA genome is translated in the chick than in the mouse cell-free system. The reason for this is not obvious, but it does not appear to be the result of nucleolytic degradation of the RNA.

**E. De Clercq:** Do I understand correctly from your talk that interferon appears to act at the level of transcription with VSV, at the level of translation with vaccinia and that it could work at either level with SV40?

**I. M. Kerr:** Taking all of the results at their face value, yes, that is the situation.

**E. De Clercq:** Might the site of action of interferon be different according to the different enzymes present in the virion?

**I. M. Kerr:** Yes, this is at least a possibility.

**H. A. Rosenthal:** Could either Dr. Kerr or Dr. De Clercq indicate what is the most recent information on the extent of purification of interferon?

**E. De Clercq:** The most "purified" interferon preparations are still relatively impure. These include preparations reported by K. H. Fantes on interferon in general (*J. gen. Physiol.* **56**, 113S, 1970) and on chick interferon (*Science, N.Y.* **163**, 1198, 1969); by W. A. Carter on mouse interferon (in Proceedings of International Meeting on Interferon and Interferon Inducers, Leuven, Belgium, Sept. 1971), as well as on mouse and human interferons (*Proc. natn. Acad. Sci. U.S.A.* **67**, 620, 1970); by K. Paucker, B. J. Berman, R. R. Golgher and D. Stanczek on mouse interferon (*J. Virol.* **5**, 145, 1970); by D. Stanczek, M. Gressnerova and K. Paucker on mouse, human and rabbit interferons (*Virology* **41**, 740, 1970). There is one inherent difficulty involved in the purification of interferons, the more highly purified they are, the less stable they become.

## NOTE ADDED IN PROOF

Further to the remarks in the introduction concerning colicin E3, it would now seem [66]* that it may be essential for the colicin to enter the bacterial cell in order for it to exert its inhibitory function. While there is no longer, therefore, an apparent precedent for an effect of this type, it remains possible that a membrane-mediated modification might be involved in interferon action.

* (See reference list)

# Nucleic Acids as Interferon Inducers

ERIK DE CLERCQ*

*Rega Institute for Medical Research,
University of Leuven, Leuven, Belgium*


Nucleic acids of both biological origin (viruses) (Table 1) and synthetic origin (homopolynucleotide pairs, alternating copolynucleotides and homopolynucleotides) (Table 2) have been shown to stimulate interferon production *in vitro* and *in vivo*.

Table 1. Nucleic acid inducers of interferon

Nucleic acids of biological origin

1. RNA
    Double-stranded RNA
        mammalian viruses: e.g. reovirus 3.
        insect viruses: e.g. cytoplasmic polyhedrosis virus.
        plant viruses: e.g. rice dwarf virus.
        mycophages: e.g. *Penicillium stoloniferum* mycophage.
        bacteriophages: e.g. MS 2 coliphage (replicative form).
        normal (mammalian) cells: e.g. rat liver cells.
    Single-stranded RNA
        double-stranded RNA formed during replication: e.g. mengo virus.
        no (detectable) double-stranded RNA formed during replication: e.g. heat-inactivated Semliki Forest virus.
2. DNA
    double-stranded RNA formed during replication: e.g. vaccinia virus.
    no (detectable) double-stranded RNA formed during replication: e.g. $T_4$ coliphage.

---

* Aangesteld Navorser of the Belgian "Nationaal Fonds voor Wetenschappelijk Onderzoek" (NFWO).

Table 2. Nucleic acid inducers of interferon

Nucleic acids of synthetic origin

1. RNA

   Double-stranded RNA
   homopolymer pairs: (poly rI).(poly rC); (poly rA).(poly rU); (poly rG). (poly rC).
   alternating copolymers: poly r(I-C); poly r(A-U); poly r(G-C).

   Single-stranded RNA
   homopolymers: poly rI, poly rG, poly rX.

2. DNA

   Double-stranded DNA
   homopolymer pairs: (poly dI).(poly dC); (poly dA).(poly dT); (poly dG).(poly dC).
   alternating copolymers: poly d(I-C); poly d(A-T).

3. Nucleic acid analogues

   thiophosphate-substituted polynucleotides: poly r($\overline{sAsU}$); poly r($\overline{sIsC}$)
   vinyl-substituted polynucleotides: (poly rI).(poly VC).

## THE TRIGGERING MOLECULE

The interferon inducing capacity of double-stranded RNA viruses (Table 1) such as reovirus type 3, MS 2 coliphage (replicative form), *Penicillium funiculosum, P. stoloniferum, P. cyaneofulvum, P. chrysogenum,* rice dwarf virus and cytoplasmic polyhedrosis virus is generally attributed to their double-stranded RNA content [3, 4, 10, 13, 50, 74, 76, 82, 94, 101, 118]. Double-stranded RNA extracted from these viruses are invariably active in inducing interferon and their interferon response peaks earlier than the interferon response obtained with the intact virus particles [13, 101, 118]. Recently, double-stranded RNA with interferon-inducing or virus-inhibiting properties has also been demonstrated in ostensibly uninfected mammalian cells, e.g. in rat liver cells by Harel and Montagnier [66] and De Maeyer et al. [38]; in rabbit kidney, chicken embryo, and HeLa cells by Kimball and Duesberg [73].

DNA viruses and single-stranded RNA viruses are also potent interferon inducers. Whether they initiate the production of interferon through formation of a double-stranded RNA intermediate is not completely clear. There are strong indications that the double-stranded RNA formed during the multiplication of mengo virus in L cells is responsible for the interferon inducing capacity of mengo virus [45, 46]. There is also suggestive evidence to believe that interferon production by DNA viruses may be mediated by a double-stranded RNA replicative form, since a viral-specific double-stranded RNA, with interferon inducing properties, has been isolated in chick cells infected with vaccinia virus by Colby and Duesberg [20].

There are, however, several conditions in which single-stranded RNA viruses do not replicate, do not initiate specific viral RNA synthesis, yet stimulate the production of interferon: temperature-sensitive mutants of Sindbis virus at non-permissive temperatures [88], temperature-sensitive mutants of Semliki Forest virus at high multiplicities [89], heat-inactivated Semliki Forest virus [60], fowl plague and Newcastle disease viruses inactivated by hydroxylamine or ultraviolet irradiation [55, 56], Newcastle disease virus in L cells treated with actinomycin D [41]. These observations have been interpreted to mean that, in some conditions, the single-stranded RNA of the input virus is a sufficient stimulus for interferon production [55, 56, 60, 89]. That single-stranded viral RNA might be capable of inducing interferon is not surprising as interferon production has been described with some synthetic single-stranded RNAs, especially when complexed to basic substances such as diethylaminoethyl (DEAE)-dextran, methylated albumin, protamine and neomycin [6, 11].

However, the inability of single-stranded RNA viruses to replicate and to produce progeny may not necessarily be regarded as inability to synthesize double-stranded RNA: e.g. chick cells infected with ultraviolet-irradiated Newcastle disease virus produce small amounts of virus-specific double-stranded RNA but do not produce infectious virus [70]. It must be said, however, that the formation of such a double-stranded virus RNA in cells infected with ultraviolet-irradiated Newcastle disease virus could not always be confirmed [56].

Recently, it has been postulated that small amounts of double-stranded RNA may be present in the virions of single-stranded RNA viruses and account for their interferon-inducing capacity (in the absence of virus replication) [52]. Thus, the interferon production by single-stranded RNA viruses can be attributed to either (1) single-stranded RNA of the input virions, or (2) "contaminating" double-stranded RNA in the input virions, or (3) double-stranded RNA formed in the infected cell during (detectable or undetectable) virus replication.

Another observation which seems to refute the likelihood of double-stranded RNA being the universal and ultimate stimulus for interferon production is that of Kleinschmidt et al. [75] who found $T_4$ coliphage capable of inducing interferon. Evidence was presented that the DNA content and not extraneous endotoxin was responsible for the interferon inducing capacity of $T_4$ coliphage. It is unlikely that $T_4$ DNA initiates interferon synthesis through mediation of a double-stranded RNA as it has not been reported to stimulate the synthesis of a double-stranded RNA in mammalian cells. The interferon inducing capacity of $T_4$ phage is probably due to the specific structural configuration of the $T_4$ DNA as it exists inside the intact phage [75].

## STRUCTURAL REQUIREMENTS

The structural requirements for a nucleic acid to be active as an inducer of interferon have been most extensively studied with synthetic polynucleotides. A variety of synthetic polynucleotides, including ribo- and deoxyribohomopolymers, homopolymer pairs and alternating copolymers (Table 2) has been shown effective in stimulating interferon production by Field *et al.* [49, 51], De Clercq and Merigan [22], and Colby and Chamberlin [19]. Their antiviral activity is greatly influenced by their molecular weight, thermal stability, resistance to degradation by nucleases and the presence of a $2'$-hydroxyl group (Table 3).

Table 3. Interferon production by synthetic polynucleotides

| Structure-function relationship |
| --- |
| 1. High molecular weight. |
| 2. Stable secondary, highly ordered, double-stranded structure. |
| 3. Resistance to degradation by nucleases. |
| 4. Presence of a $2'$-hydroxyl group. |

The influence of the molecular sizes of the individual homopolymers polyriboinosinic acid and polyribocytidylic acid on the activity of the double-stranded complex (poly rI).(poly rC) is shown in Table 4 (De Clercq and Merigan, unpublished data, 1970). A reduction in the molecular weight of the constituent components was accompanied by a gradually decreased capacity to induce interferon in the rabbit. The capacity of (poly rI).(poly rC) to induce interferon and resistance to virus infection in cell culture was markedly less affected by molecular weight reduction. Similar results have been obtained in other studies by Tytell *et al.* [119] and Lampson *et al.* [78]. With preparations of high molecular weight (poly rI).(poly rC), which were exposed to sonic radiation, significant decrease in activity was noted in (poly rI).(poly rC) fractions with approximate molecular weight less than $1.2 \times 10^5$ [78]. The antiviral activity of (poly rI).(poly rC) was more dependent upon maintaining a high molecular weight of poly rI than of poly rC; activity was not markedly reduced unless the molecular weight of poly rI was below $1.9 \times 10^5$ and of poly rC below $2.3 \times 10^4$ [119].

Double-stranded RNA complexes need Tm (thermal stability) values higher than $60°C$ (calculated for 0.15 M $Na^+$) to be fully active as interferon inducers and rapidly decrease in activity if they melt out at lower temperatures [22, 31, 35]. Since thermal stability represents a valuable measure of the overall stability of double-stranded polyribonucleotides [43], polymers with Tm values higher than 60°C may also be considered to be the most stable ones at 37°C. It seems as though complementary strands have to form a stable secondary, highly ordered double-stranded structure in order to be effective interferon inducers.

A high molecular weight and a stable secondary structure may influence the antiviral activity of the polynucleotide by regulating the interaction of the polynucleotide with the cell, by promoting its affinity for the cellular receptor site for interferon production, or, simply, by increasing its resistance to premature enzymatic degradation. More stable and higher molecular weight polynucleotide complexes which are most active in inducing interferon are also

Table 4. Influence of the molecular weight of the constituent components on the antiviral activity of (poly rI). (poly rC)

| Molecular weight ($S_{20,W}$) | | In vitro (human skin fibroblasts) | | In vivo (rabbits) |
|---|---|---|---|---|
| Poly rI | Poly rC | Cellular resistance to virus infection M.I.C.* ($\mu$g/ml) | Interferon production (units/4 ml)† | Interferon production (units/4 ml)‡ |
| 2.50 | 3.08 | $4 \cdot 10^{-5}$ | 45 | 280 |
| 4.39 | 3.88 | $10^{-5}$ | 160 | 320 |
| 6.13 | 5.94 | $10^{-5}$ | 300 | 560 |
| 7.94 | 8.20 | $10^{-5}$ | 260 | n.t.§ |
| 10.6 | 10.5 | $10^{-5}$ | 280 | 1280 |
| 12.5 | 13.2 | $10^{-5}$ | 280 | 1800 |

\* Minimal inhibitory concentration: polymer concentration required to reduce (vesicular stomatitis) virus plaque formation by 50%. Different polymer concentrations were prepared in minimal Eagle's medium, preincubated at 37°C and then applied to the cells.

† Preheated (37°C) (poly rI).(poly rC) exposed to the cells for three hours at 0.4 $\mu$g/ml; interferon production measured in the supernatant fluid at the end of the subsequent six to eight hour incubation period.

‡ Serum interferon titres measured two hours after intravenous injection of 2 $\mu$g of the polymer.

§ n.t.: not tested.

more resistant to nucleases. There are exceptions, however. For example, poly r(A-U) is more active and more stable than (poly rA).(poly rU), yet it is more sensitive to pancreatic ribonuclease [19]. Several systems have been developed to potentiate the production of interferon by polynucleotides: e.g. substitution of thiophosphate for phosphate in the polymer backbone, preincubation of the polymer at 37°C in tissue culture medium, and addition of polycations such as DEAE-dextran (to the polymer or to the cells). Substitution of thiophosphate for phosphate as well as preincubation at 37°C increased the interferon inducing capacity of polyribonucleotides and their resistance to degradation by ribonuclease in parallel [30, 32, 35, 36]. An increased resistance to degradation by nuclease might also explain the increased antiviral activity of (poly rI).(poly rC) in the presence of DEAE-dextran; and it might also help to explain the high antiviral activity of (poly rI).(poly VC). (Complex of polyriboinosinic acid and poly(1-vinylcytosine): Pitha and Pitha [100].)

Although polydeoxyribonucleotides have a similar or greater molecular weight and structural stability, they are significantly less active than their RNA counterparts [19, 31, 32, 124]. DNA-RNA hybrids of either synthetic or biological origin [19, 21, 124] are as impotent as DNA polymers. The differences in activity between polyribo- and polydeoxyribonucleotides could be related to differences in configuration (A helix vs B helix), but this assumption is unable to explain why DNA-RNA hybrids are less effective than double-stranded RNA, as both species occur in the A helix configuration. It would appear, therefore, that the $2'$-hydroxyl groups have a specific effect on the interferon inducing capacity of polynucleotides. In order to assign a definitive role to the $2'$-hydroxyl groups, it should be worthwhile to determine the antiviral activity of polynucleotides in which the $2'$-hydroxyls have been replaced by other radicals, e.g. $O$-methyl groups or halogens. Preliminary data of experiments carried out in collaboration with Drs D. Shugar and B. Zmudska of Warszawa (Poland) indicate that substitution of $2'$-$O$-methyl for $2'$-OH in poly rC results in a dramatic reduction of the antiviral activity of (poly rI).(poly rC).

In a recent report, Pitha and Pitha [100] tend to de-emphasize the need for stringent structural requirements: they postulate that the antiviral activity of polynucleotides is determined by the ease of cellular uptake of the polynucleotide. This conclusion is based on the high antiviral activity of (poly rI).(poly VC) and (poly rI).(poly rC)-DEAE–Dextran complexes. As compared to (poly rI).(poly rC), (poly rI).(poly VC) and (poly rI).(poly rC)-DEAE-dextran complexes have a lower charge/mass ratio and occur in a more aggregated form. These two characteristics enhance the cellular uptake of the polynucleotide and might be responsible for the increased antiviral activity, according to Pitha and Pitha [100].

## THE TRIGGERING PROCESS

How do the requirements for a high molecular weight, a stable secondary structure, resistance to degradation by nucleases and the presence of $2'$-hydroxyl groups, affect the interaction of the polynucleotide with the cell and the process of interferon formation? Polynucleotides attach very rapidly to the cells (within seconds) [De Clercq, unpublished data, 1971]; the rate of cell-binding and the amount of polymer bound to the cells are markedly increased if the polymers are preincubated at 37°C [36, 37].

Although DNA polymers are less effective in stimulating interferon production, they attach more rapidly to the cells than RNA polymers [37]. Thus, differences in antiviral activity among polyribo- and polydeoxyribonucleotides cannot be explained by differences in the rate of cell-binding. Nor could they be explained by differences in persistance of polymer at the outer cell membrane (determined by measuring the amount of cell-bound polymer that was released by nuclease treatment following an initial incubation of the cells with the polynucleotide for 1 hour). The persistence of the polymer at the cell surface appeared to be inversely correlated to the antiviral activity with a variety of RNA and DNA polymers tested [37]. These findings may be interpreted as follows: (A) only a minute quantity of cell-bound polymer is directly involved in the interferon inducing process; interferon production is initiated (B) during the interaction of the polymer with a specific receptor site at the cell surface, or (C) during penetration of the polymer into the cell, or (D) during the uptake and degradation of the polymer in phagosomes. The critical step for interferon production cannot be identified with the reutilization of degraded polymer material into host RNA synthesis as such utilization is completely blocked by actinomycin D and cytidine under conditions where the interferon response is not affected [8].

It has recently been demonstrated that mouse L cell-bound (poly rI).(poly rC) is capable of stimulating the production of interferon in rabbit kidney cell cultures [28]. These findings suggest that (poly rI). (poly rC) does not have to penetrate into the cell in order to elicit the interferon response.

Studies on the interaction of polynucleotides with the cell have revealed that the antiviral activity of (poly rI) . (poly rC) in cell cultures can be restored if both homopolymers are administered successively [27]. Successive administration of the individual homopolymers in the order poly rI, poly rC resulted in a greater antiviral activity than that resulting from addition of (poly rI) . (poly rC) itself. The priming effect of poly rI on the antiviral activity of poly rC was observed with either VSV (vesicular stomatitis virus) or vaccinia virus in a variety of cell cultures [continuous cell lines such as human skin fibroblasts (HSF), rabbit kidney (RK 13), mouse L 929 cells, HeLa cells and human kidney (HK) cells; and primary mouse embryo fibroblasts (MEF) and primary rabbit kidney (PRK) cells] (Fig. 1). The antiviral activity of (poly rI) . (poly rC) was only partially restored if the individual homopolymers were administered in the order poly rC, poly rI (with 30 min interval between the administration of both polymers). In PRK cell cultures, however, sequential administration of poly rC and poly rI gave a greater antiviral activity than the (poly rI) . (poly rC) complex (Fig. 1).

The extent of the priming effect of poly rI on the antiviral activity of poly rC did not depend on the time of contact of the cells with poly rI, no matter whether poly rI had been exposed to the cells for 10 seconds or 24 hours (Fig. 2). Similarly, the priming effect of poly rC did not change with the time of exposure of the cells to poly rC; in RK 13 cells, however, the priming activity of poly rC wore off quickly and disappeared if poly rC was incubated with the cells for more than one hour (Fig. 2).

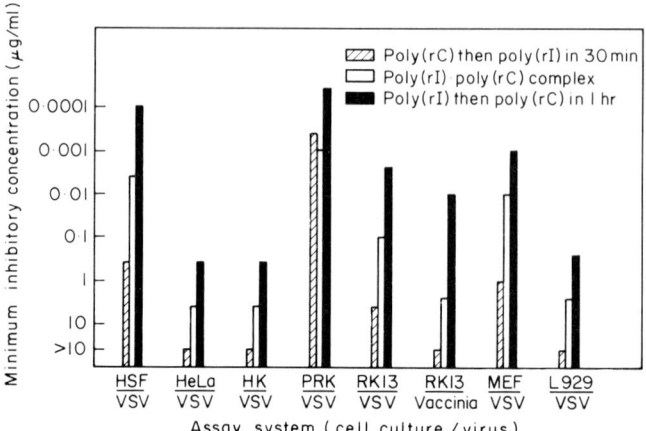

**Figure 1.** Antiviral activity of poly (rI) and poly (rC) in different cell culture/virus assay systems. Poly (rI) and poly (rC) were administered in sequence or in complex form. The minimum inhibitory concentration corresponds to the concentration of polymer [either poly (rI) . poly (rC) or poly (rI), if poly (rI) is added second, or poly (rC), if poly (rC) is added second] required to reduce virus plaque formation by 50%. In cells primed with poly (rI), poly (rC) was added 1 hour after poly (rI); in cells primed with poly (rC), poly (rI) was added 30 min after poly (rC). The first polymer was applied at 10 μg/ml; the second polymer was applied at different concentrations (from $10^{-5}$ to 10 μg/ml).

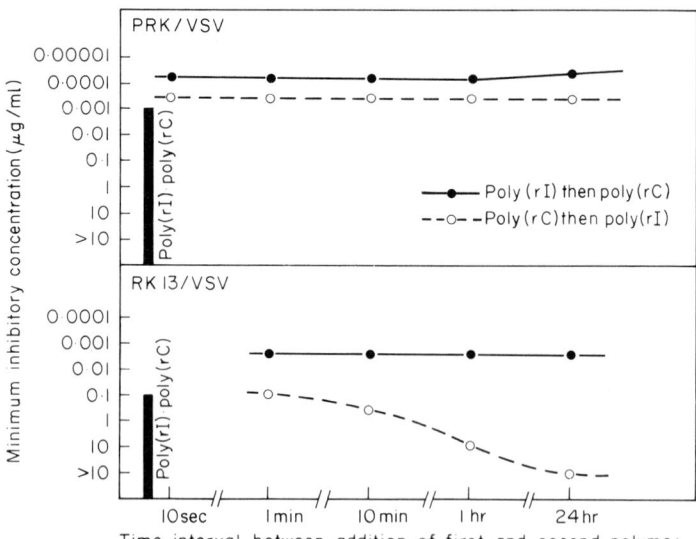

**Figure 2.** Antiviral activity of poly (rI) in cells primed with poly (rC) for different times and of poly (rC) in cells primed with poly (rI) for different times. The antiviral activity was measured in two different assay systems: PRK/VSV and RK 13/VSV. The minimum inhibitory concentration corresponds to the concentration of polymer [either poly (rC) . poly (rC) or poly (rI), if poly (rI) is added second, or poly (rC), if poly (rC) is added second] required to reduce virus plaque formation by 50%.

Two hypotheses can be proposed to account for the antiviral activity of the successively administered homopolymers: (1) poly rI and poly rC form a double-stranded complex at the outer cell membrane or within the cell; the antiviral effect of either poly rI in poly rC-primed cells or of poly rC in poly rI-primed cells is due to the newly formed (poly rI) . (poly rC) complex; or (2) poly rI and poly rC do not anneal but act independently on the cell; the first homopolymer causes an effect required for the antiviral activity of the second polymer. The following observations (De Clercq and De Somer, to be published) indicate that poly rI and poly rC do not act independently but reunite at the cellular level:

1° Poly rI did not prime for the antiviral activity of non-complementary homopolynucleotides (e.g. poly rU, poly rG).
2° The priming effect of poly rI was significantly reduced if the poly rI-primed cells were treated with $T_1$ ribonuclease or DEAE-dextran before addition of poly rC.
3° The priming effect of poly rC was significantly reduced if the poly rC-primed cells were treated with pancreatic ribonuclease or DEAE-dextran before addition of poly rI.

$T_1$ ribonuclease and pancreatic ribonuclease were able to reverse the priming effect of either poly rI or poly rC, even if the primer molecule had been incubated on the cells for 24 hours.

## *IN VIVO* ACTIVITY

Favourable results have been reported with double-stranded polyribonucleotides such as (poly rI) . (poly rC) in the prophylaxis of a wide variety of experimental virus infections in animals. Many of these animal model systems resemble acute virus infections in man, in that the route of infection and pathogenesis of the disease are similar (Table 5). Even if treatment was delayed until three or five days after virus inoculation, at the time that high virus titres were reached in the target organs or clinical symptoms appeared, (poly rI) . (poly rC) was still effective against herpes simplex virus keratoconjunctivitis in rabbits [98], vesicular stomatitis virus encephalitis [23, 34], pneumonia virus [93], and Semliki Forest virus encephalitis (130) in mice. Evidence has been presented that, at least in one experimental model (vesicular stomatitis virus encephalitis), the whole protective activity of (poly rI) . (poly rC) is due to interferon production: treatment with exogenous interferon in amounts which closely mimicked the levels of circulating interferon produced endogenously by (poly rI) . (poly rC) gave essentially the same antiviral protection as the corresponding dose of (poly rI) . (poly rC) [34].

Protective effects with (poly rI) . (poly rC) have also been obtained in experimental infections with protozoa (*Plasmodium berghei* in mice: Jahiel *et al.*

[71]; *Trypanosoma congolense* in mice: Herman and Baron [68]; *Toxoplasma gondii* in rabbits: Chowchuvech *et al.* [18]; Oh and O'Connor [95]), *Chlamydia trachoma* in rabbits: Oh *et al.* [96]), bacteria (*Escherichia coli* in mice: Weinstein *et al.* [125]; *Listeria monocytogenes* in mice: Remington and Merigan [106]) and with fungi (*Cryptococcus neoformans* in mice: Regelson and

Table 5. Antiviral activity of (poly rI) . (poly rC) in animal model systems related to acute natural virus infections in man

| Animal model system | Human parallel | References |
|---|---|---|
| Herpes simplex virus keratoconjunctivitis (rabbits) | Herpes simplex keratitis | (a) |
| Herpes simplex virus encephalitis (mice) | Herpes encephalitis | (b) |
| Vesicular stomatitis virus encephalitis (mice) | Herpes or enterovirus encephalitis | (c) |
| Rabies virus encephalomyelitis (rabbits) | Rabies | (d) |
| Columbia SK virus encephalitis (mice) | Enterovirus encephalitis | (e) |
| Semliki Forest virus encephalitis (mice) | Arbovirus encephalitis | (f) |
| Japanese B virus encephalitis (mice) | Arbovirus encephalitis | (g) |
| Vaccinia virus tail lesions (mice) | Small pox (viremic phase) | (h) |
| Influenza A and B virus respiratory infection (mice, baboons) | Influenza (and other respiratory infections) | (i) |
| Parainfluenza virus type 1 (Sendai) and type 3 respiratory infection (mice, hamsters) | Influenza (and other respiratory infections) | (j) |
| Rhinovirus type 13 virus respiratory infection (man) | Cold (upper respiratory infections) | (k) |

(a) Park and Baron [98];
(b) Catalano and Baron [16]; Hamilton *et al.* [64];
(c) De Clercq and Merigan [23]; De Clercq *et al.* [34];
(d) Fenje and Postic [47, 48];
(e) Lindh *et al.* [86]; Nemes *et al.* [93];
(f) Finter [53]; Worthington and Baron [130];
(g) Postic and Sather [102];
(h) Nemes *et al.* [93];
(i) Gerone *et al.* [58]; Heberling and Kalter [67]; Lindh *et al.* [86]; Nemes *et al.* [93];
(j) Gerone *et al.* [58]; Lindh *et al.* [86]; Nemes *et al.* [93]; Renis [107];
(k) Hill *et al.* [69].

Munson, [103]). Some of these agents are also sensitive to the action of interferon: *Chlamydia* (trachoma: Hanna *et al.* [65]), protozoa (*Toxoplasma gondii*: Remington and Merigan [105]), bacteria (*Shigella flexneri*: Gober *et al.* [59]). The inhibitory effect of interferon on *Chlamydia*, protozoa and bacteria has essentially been demonstrated in cell cultures. It is not clear yet whether interferon mediates the inhibitory effects of (poly rI) . (poly rC) *in vivo*.

However, there is suggestive evidence that the protective effect of (poly rI) . (poly rC) in *Trypanosoma congolense*-infected mice is mediated by an enhancement of the immune responsiveness and not by interferon production [68].

Furthermore, (poly rI) . (poly rC) inhibits the growth of a variety of tumours, whether or not they are known to depend on continued virus replication (viral induced, chemically induced, transplanted and spontaneous tumours) (Table 6). Although interferon itself is able to suppress the growth of these tumours, and in

Table 6. Antitumour activity of (poly rI) . (poly rC)

I.  Virus-induced tumours:
    Moloney sarcoma (mice)              Sarma et al. [111]; Weinstein et al. [126]
    Rous sarcoma (chicken)              Nemes et al. [93]
    Friend leukaemia (mice)             Larson et al. [79]
    Adenovirus type 12 (hamsters)       Larson et al. [80]
    Polyoma (rats)                      Vandeputte et al. [120]
    SV 40 (hamsters)                    Larson et al. [81]
    Shope fibroma (rabbits)             Friedman-Kien and Vilcek [54]

II. Chemically induced tumours:
    Dimethylbenzanthracene-induced skin tumour (mice): Gelboin and Levy [57]

III. Transplanted tumours:
    L 1210 leukaemia (mice)             Zeleznick and Bhuyan [135]
    Reticulum cell sarcoma, etc.
    (mice)                              Levy et al. [85]
    RC 19 tumour (mice)                 Gresser and Bourali [61]
    Malignant melanoma (mice)           Bart and Kopf [7]
    AKR/J leukaemia (mice)              Meier et al. [92]
    Ehrlich ascites tumour (mice)       Rhim and Huebner [108]
    Moloney sarcoma (mice)              Weinstein et al. [126]

IV. Spontaneous tumours:
    Mammary carcinoma (mice)            Came and Moore [14]
    Acute lymphoblastic leukaemia
    (man)                               Mathé et al. [91]

the continued virus replication may be critical in maintaining progression of the malignancy [62, 127], and may also inhibit tumours which do not contain infectious oncogenic viruses [61], it is difficult to assess the role of interferon in the anti-tumour effect observed with (poly rI) . (poly rC) in so many different test conditions (Table 6). Weinstein et al. [126] did not find a correlation between the interferon response to (poly rI) . (poly rC) and the anti-tumour activity of (poly rI) . (poly rC) in different mouse strains infected with Moloney sarcoma virus: only two out of four strains were protected by (poly rI) . (poly rC) against Moloney sarcoma virus-induced tumour formation, although all strains developed equivalent peak concentrations of interferon following injection of (poly rI).(poly rC). These data do not necessarily exclude the possibility that the antitumour effect of (poly rI).(poly rC) is mediated by interferon production, for (Moloney sarcoma virus-induced) tumour formation might be more sensitive to interferon in some mouse strains than in others. In our studies, exogenous interferon, administered in amounts equivalent to those induced endogenously by (poly rI).(poly rC) gave essentially the same protection against Moloney sarcoma virus-induced tumours as (poly rI).(poly rC), suggesting that the whole anti-tumour effect of (poly rI).(poly rC) in the conditions used [single dose of (poly rI).(poly rC) injected before virus infection] is accounted for by interferon production [29].

Double-stranded RNAs have profound effects on host defence mechanisms; they stimulate phagocytosis [104], cell mediated immunity [15, 117] and humoral immunity [128, 129], are pyrogenic [87] and may also exert a direct chemotherapeutic effect on tumours [84]. Depending on the test conditions, these effects might prove as important as interferon itself in the activity of (poly rI).(poly rC) against tumour formation and bacterial, protozoal and fungal infections.

## POTENTIATION OF THE INTERFERON RESPONSE

Several methods have been developed for potentiating the interferon stimulating activity of polynucleotides (Table 7).

Table 7. Interferon inducing capacity of synthetic polynucleotides increased through

| In vitro | In vivo |
|---|---|
| (1) addition of low- or high-molecular weight polycations (e.g. DEAE-dextran, neomycin) | (1) addition of high molecular weight polycations (e.g. DEAE-dextran, polylysine) |
| (2) substitution of thiophosphate for phosphate [e.g. poly r(sAsU), poly r(sIsC)] | (2) substitution of thiophosphate for phosphate [e.g. poly r(sAsU), poly r(sIsC)] |
| (3) thermal activation (preincubation of the polymer at 37°C in tissue culture medium) | (3) addition of metabolic inhibitors (e.g. cycloheximide) |
| (4) addition of metabolic inhibitors (e.g. actinomycin D, cycloheximide) | (4) intraperitoneal injection of Freund's adjuvant |
| (5) priming with interferon | (5) intravenous injection of lead acetate |
| (6) priming with (poly rI).(poly rC) | |
| (7) separate administration of complementary homopolynucleotides (e.g. poly rI followed by poly rC) | |

Addition of low-molecular-weight polycations such as polyamine antibiotics (neomycin, streptomycin) increases interferon production by (poly rI) . (poly rC) *in vitro* [11, 77] but not *in vivo* [77]. Addition of high-molecular-weight polycations such as DEAE-dextran or poly-lysine increases the interferon response to (poly rI) . (poly rC) both *in vitro* [8, 19, 26, 39, 42, 44, 116, 124] and *in vivo* [40, 53, 109]. These polybasic substances may act by promoting interaction of the polynucleotide with the cell and/or increasing its resistance to nuclease degradation within the cell or at the cell surface.

In view of the findings of Nordlund *et al.* [94a] and Stern [111a] that double-stranded polyribonucleotides such as (poly rI) . (poly rC) are rapidly degraded by nucleases present in the serum, any system designed to increase the resistance of double-stranded polyribonucleotides to nuclease degradation might have clinical significance. Substitution of thiophosphate for phosphate and preincubation of the polymer at 37°C enhanced the interferon inducing capacity of the alternating copolymers poly r(A-U) and poly r(I-C), apparently as a consequence of an increased resistance to ribonuclease [30, 31, 32, 35, 36]. This potentiation was demonstrated *in vitro* and *in vivo* with the thiophosphate-substituted polynucleotides but only *in vitro* with the heated polymers.

The effect of inhibitors of RNA and protein synthesis on the interferon production by nucleic acids (either viruses or synthetic polynucleotides) is rather complex. It depends on both dosage and timing of administration of the metabolic inhibitor. In general metabolic inhibitors (actinomycin D, cycloheximide) inhibit interferon production during its early phase but enhance interferon production during its declining phase (*in vitro*: Vilcek *et al.* [123]; Vilcek [121]; Vilcek and Ng, [122]; Tan *et al.* [114, 115]; *iv vivo*: Youngner and Hallum [132]; De Clercq and Merigan [24]). To the degree that the action of cycloheximide can be interpreted to block protein synthesis, it may be concluded that new protein synthesis is required for both the initiation and termination of the interferon production. Cycloheximide might interfere with the turnoff of interferon production by blocking the synthesis of a protein which inhibits the formation (or action) of interferon. Alternatively, cycloheximide might enhance interferon production by other mechanisms e.g. by stabilizing interferon against inactivation, or by stimulating recirculation of interferon.

There are two other methods potentiating interferon production by (poly rI) . (poly rC) *in vivo*: first, intraperitoneal injection of Freund's adjuvant (complete or incomplete Freund's adjuvant) or mineral oil [33], and second, intravenous injection of lead acetate [113]. An increased uptake of the interferon inducer by the reticuloendothelial cells has been proposed to account for the effect of Freund's adjuvant.

The interferon response to (poly rI) . (poly rC) in cell cultures pretreated with interferon has been studied by several authors; a decreased interferon response was recorded in some studies by Youngner and Hallum [133], an increased interferon response in other studies by Rosztoczy and Mécs [110], Bausek and Merigan [9], Margolis et al. [90], Barmak and Vilcek [5]. Stewart et al. [112] found a decreased interferon response in interferon-treated cells if (poly rI). (poly rC) was employed at an optimal concentration and increased interferon yields if (poly rI). (poly rC) was used at a suboptimal concentration. An increased interferon response to (poly rI). (poly rC) has also been obtained in human skin fibroblast cultures which had been previously exposed to (poly rI). (poly rC) [12]. These results suggest that both interferon and (poly rI). (poly rC) can be applied successfully to sensitize the cells for the interferon producing capacity of polynucleotides.

As discussed above, priming of the cells with poly rI, followed by treatment with poly rC, resulted in a greater antiviral effect than direct exposure of the cells to (poly rI). (poly rC). This priming effect was obtained in several cell cultures challenged with different viruses. *In vivo*, the antiviral activity of (poly rI). (poly rC) was only partially restored if poly rI and poly rC were injected separately [27].

## TOXICITY

In addition to the broad range of effects on host defence mechanisms (described above), (poly rI). (poly rC), and probably all double-stranded RNAs, have various side effects disturbingly similar to those observed with bacterial endotoxin [25]. (Poly rI). (poly rC) is lethal for mice at relatively low dosages, provokes a local Shwartzman phenomenon, is highly pyrogenic and embryotoxic in rabbits and induces a runt-like disease in mice and rats; toxic effects have also been noted on the rabbit eye (inflammation, lens opacity), the chick cerebellum (cerebellar ataxia, due to damage of the endothelial cells of the small blood vessels) and on mouse haemopoietic stem cells (1, 2, 72, 83, 87, 97, 131). The toxicity of (poly rI). (poly rC) was associated with the double-strandedness of the compound and did not occur with the single homopolymers poly rI and poly rC [72, 87]. As shown with endotoxin, the lethal action of (poly rI). (poly rC) was significantly enhanced in adrenalectomized rats [134] and in mice treated with lead acetate [1] or actinomycin D [99].

These toxic properties have most extensively been explored with (poly rI). (poly rC) and the question may be raised whether they are inseparably linked to the antiviral activity of double-stranded RNAs. Although no comparative studies have been carried out on the therapeutic ratios (ratio of maximum tolerated dose to minimum effective dose) of different polynucleotides in different systems, there are at least some indications that toxicity and antiviral activity of synthetic polynucleotides are closely associated phenomena (Table 8). Simultaneous injection of lead acetate increased both lethality and interferon production by (poly rI). (poly rC), so that the toxicity

Table 8. Toxicity to antiviral activity ratios of double-stranded polyribonucleotides in the mouse*

|  | Exp. 1 | Exp. 2 | | |
|---|---|---|---|---|
|  | (Poly rI) · (poly rC) | (Poly rI) · (poly rC) | Poly r(A-U) | Poly r(s̄As̄U) |
| Normal mice |  |  |  |  |
| Minimum effective dose (A) | 0.1 | 0.2 | 50 | 15 |
| Toxic dose LD$_{50}$ (B) | 300 | 200 | — | — |
| Ratio B/A | 3000 | 1000 | — | — |
| Lead-acetate-treated mice† |  |  |  |  |
| Minimum effective dose (A') | 0.002 | — | — | — |
| Toxic dose LD$_{50}$ (B') | 2 | 0.4 | 50 | 15 |
| Ratio B'/A' | 1000 | — | — | — |
| Ratio B'/A | 20 | 2 | 1 | 1 |

* Dosage levels expressed in μg per mouse. Male NMRI mice (10-12 g) used in Exp. 1; Female Swiss-Webster mice (10-12 g) used in Exp. 2. All polynucleotides were injected intravenously.
† Lead acetate injected intravenously at 1 mg per mouse immediately before the polynucleotide.
A,A' Dose of polynucleotide which stimulated production of circa 40-50 units of interferon per 4 ml serum.
B,B' Dose of polynucleotide which was lethal for 50% of the mice.

to antiviral activity ratio did not change significantly. Similarly, introduction of sulphur in the phosphate linkages of poly r(A-U) increased its antiviral activity and lethal effect in parallel, so that identical toxicity to anti-viral activity ratios (measured by comparing the toxic dose in lead acetate-treated mice to the minimum effective dose in normal mice) were recorded with poly r(A-U) and its thiophosphate analogue poly r(sAsU) (De Clercq, Eckstein, Sternbach and Merigan, 1971, to be published).

## SUMMARY

Nucleic acids of both biological and synthetic origin are effective inducers of interferon *in vitro* and *in vivo*. Although both single- and double-stranded RNA and DNA polymers are capable of stimulating interferon production, double-stranded RNAs are the most active ones. The interferon inducing capacity of viruses is generally ascribed to their double-stranded RNA content or their capacity to induce a double-stranded RNA replicative form in the host cell. There are, however, some conditions in which (DNA or single-stranded RNA) viruses do not replicate, yet stimulate the production of interferon.

The interferon inducing capacity of synthetic polynucleotides depends on their molecular weight, structural stability, resistance to nuclease degradation and the presence of $2'$-hydroxyl groups. It is not completely clear how these structural requirements affect the interaction of the polynucleotide with the cell and the (hypothetical) triggering or receptor site for interferon production. There is suggestive evidence, however, that this triggering site may be situated at the outer cell membrane. It has further been established that the individual polymers of a double-stranded homopolymer complex [e.g. (poly rI) . (poly rC)] might be administered separately in order to elicit the interferon response. If added successively to cell cultures, the homopolymers poly rI and poly rC do not act independently but reunite at the cellular level.

Double-stranded RNAs not only effective interferon inducers, they have profound effects on different host defence mechanisms, including an adjuvant effect on cellular and humoral immune responses. These effects might be as important as interferon production in the *in vivo* activity of double-stranded RNA against non-viral infections and neoplasms.

Several systems have been reported to potentiate the interferon response to synthetic polynucleotides either *in vitro* or *in vivo*: addition of polybasic compounds (e.g. DEAE-dextran, neomycin, polylysine), substitution of thiophosphate for phosphate, preincubation of the polymers at $37^\circ$C in tissue culture medium, addition of metabolic inhibitors (e.g. actinomycin D, cycloheximide), injection of Freund's adjuvant or lead acetate, priming of cells with interferon or interferon inducers and separate administration of complementary homopolynucleotides (e.g. poly rI followed by poly rC). There are indications that at least in some conditions (injection of lead acetate, substitution of thiophosphate for phosphate) antiviral activity and toxicity are increased in parallel.

REFERENCES

1. Absher, M. and Stinebring, W. R. (1969). *Nature, Lond.* **223**, 715.
2. Adamson, R. H. and Fabro, S. (1969). *Nature, Lond.* **223**, 718.
3. Banks, G. T., Buck, K. W., Chain, E. B., Himmelweit, F., Marks, J. E., Tyler, J. M., Hollings, M., Last, F. T. and Stone, O. M. (1968). *Nature, Lond.* **218**, 542.
4. Banks, G. T., Buck, K. W., Chain, E. B., Darbyshire, J. E. and Himmelweit, F. (1969). *Nature, Lond.* **223**, 155.
5. Barmak, S. and Vilcek, J. (1971). *Bact. Proc.* **196**.
6. Baron, S., Bogomolova, N. N., Billiau, A., Levy, H. B., Buckler, C. E., Stern, R. and Naylor, R. (1969). *Proc. natn. Acad. Sci. U.S.A.* **64**, 67.
7. Bart, R. S. and Kopf, A. W. (1969). *Nature, Lond.* **224**, 372.
8. Bausek, G. H. and Merigan, T. C. (1969). *Virology* **39**, 491.
9. Bausek, G. H. and Merigan, T. C. (1970). *Proc. Soc. exp. Biol. Med.* **134**, 672.
10. Bessell, C. J., Bolling, N. J., Fantes, K. H., Laursen, A. C., Newcomb, J. M., Pamplin, P. R. and Sutherland, E. S. (1971). Personal communication.
11. Billiau, A., Buckler, C. E., Dianzani, F., Uhlendorf, C. and Baron, S. (1969). *Proc. Soc. exp. Biol. Med.* **132**, 790.
12. Billiau, A., Van den Berghe, H. and De Somer, P. (1972). *J. gen. Virol.* in press.
13. Buck, K. W., Chain, E. B. and Himmelweit, F. (1971). *J. gen. Virol.* **12**, 131.
14. Came, P. E. and Moore, D. H. (1971). *Proc. Soc. exp. Biol. Med.* **137**, 304.
15. Cantor, H., Asofsky, R. and Levy, H. B. (1970). *J. Immun.* **104**, 1035.
16. Catalano, L. W., Jr. and Baron, S. (1970). *Proc. Soc. exp. Biol. Med.* **133**, 684.
17. Chamberlin, M., Baldwin, R. L. and Berg, P. (1963). *J. molec. Biol.* **7**, 334.
18. Chowchuvech, E., Weissenbacher, M., Schmunis, G., Sawicki, L., Galin, M. A. and Baron, S. (1971). Personal communication.
19. Colby, C. and Chamberlin, M. J. (1969). *Proc. natn. Acad. Sci. U.S.A.* **63**, 160.
20. Colby, C. and Duesberg, P. H. (1969). *Nature, Lond.* **222**, 940.
21. Colby, C., Stollar, B. D. and Simon, M. I. (1971). *Nature New Biology, Lond.* **229**, 172.
22. De Clercq, E. and Merigan, T. C. (1969a). *Nature, Lond.* **222**, 1148.
23. De Clercq, E. and Merigan, T. C. (1969b). *J. gen. Virol.* **5**, 359.
24. De Clercq, E. and Merigan, T. C. (1970a). *Virology* **42**, 799.
25. De Clercq, E. and Merigan, T. C. (1970b). *Archs intern. Med.* **126**, 94.
26. De Clercq, E. and Merigan, T. C. (1971). *J. gen. Virol.* **10**, 125.
27. De Clercq, E. and De Somer, P. (1971a). *Science, N.Y.* **173**, 260.
28. De Clercq, E. and De Somer, P. (1971b). Submitted for publication.
29. De Clercq, E. and De Somer, P. (1971c). *J. natn. Cancer Inst.* (in press).

30. De Clercq, E., Eckstein, F. and Merigan, T. C. (1969). *Science, N.Y.* **165**, 1137.
31. De Clercq, E., Eckstein, F. and Merigan, T. C. (1970a). *Ann. N.Y. Acad. Sci.* **173**, 444.
32. De Clercq, E., Wells, R. D. and Merigan, T. C. (1970b). *Nature, Lond.* **226**, 364.
33. De Clercq, E., Nuwer, M. R. and Merigan, T. C. (1970c). *Infect. Immunity* **2**, 69.
34. De Clercq, E., Nuwer, M. R. and Merigan, T. C. (1970d). *J. Clin. Invest.* **49**, 1565.
35. De Clercq, E., Eckstein, F., Sternbach, H. and Merigan, T. C. (1970e). *Virology* **42**, 421.
36. De Clercq, E., Wells, R. D., Grant, R. C. and Merigan, T. C. (1971). *J. molec. Biol.* **56**, 83.
37. De Clercq, E., Wells, R. D. and Merigan, T. C. (1972). *Virology.* (in press).
38. De Maeyer, E., De Maeyer-Guignard, J. and Montagnier, L. (1971). *Nature New Biology, Lond.* **229**, 109.
39. Dianzani, F., Cantagalli, P., Gagnoni, S. and Rita, G. (1968). *Proc. Soc. exp. Biol. Med.* **128**, 708.
40. Dianzani, F., Rita, G., Cantagalli, P. and Gagnoni, S. (1969). *J. Immun.* **102**, 24.
41. Dianzani, F., Gagnoni, S., Buckler, C. E. and Baron, S. (1970). *Proc. Soc. exp. Biol. Med.* **133**, 324.
42. Dianzani, F., Baron, S., Buckler, C. E. and Levy, H. B. (1971). *Proc. Soc. exp. Biol. Med.* **136**, 1111.
43. Doty, P., Boedtker, H., Fresco, J. R., Haselkorn, R. and Litt, M. (1959). *Proc. natn. Acad. Sci. U.S.A.*, **45**, 482.
44. Falcoff, E. and Perez-Bercoff, R. (1969). *Biochim. biophys. Acta* **174**, 108.
45. Falcoff, R. and Falcoff, E. (1969). *Biochim. biophys. Acta* **182**, 501.
46. Falcoff, R. and Falcoff, E. (1970). *Biochim. biophys. Acta* **199**, 147.
47. Fenje, P. and Postic, B. (1970). *Nature, Lond.* **226**, 171.
48. Fenje, P. and Postic, B. (1971). *J. infect. Dis.* **123**, 426.
49. Field, A. K., Tytell, A. A., Lampson, G. P. and Hilleman, M. R. (1967a). *Proc. natn. Acad. Sci. U.S.A.*, **58**, 1004.
50. Field, A. K., Lampson, G. P., Tytell, A. A., Nemes, M. M. and Hilleman, M. R. (1967b). *Proc. natn. Acad. Sci. U.S.A.*, **58**, 2102.
51. Field, A. K., Tytell, A. A., Lampson, G. P. and Hilleman, M. R. (1968). *Proc. natn. Acad. Sci. U.S.A.*, **61**, 340.
52. Field, A. K., Lampson, G. P., Tytell, A. A., Nemes, M. M. and Hilleman, M. R. (1971). *Bact. Proc.* **195**.
53. Finter, N. B. (1970). Colloques de l'Institut National de la Santé et de la Recherche Médicale, no. 6, "L'Interféron" p. 325.
54. Friedman-Kien, A. E. and Vilcek, J. (1970). Proceedings Vth International Congress of Infect. Diseases, Vienna, Aug. 31-Sept. 5, p. 239.
55. Gandhi, S. S. and Burke, D. C. (1970). *J. gen. Virol.* **6**, 95.
56. Gandhi, S. S., Burke, D. C. and Scholtissek, C. (1970). *J. gen. Virol.* **9**, 97.
57. Gelboin, H. V. and Levy, H. B. (1970). *Science, N.Y.* **167**, 205.
58. Gerone, P. J., Hill, D. A., Appell, L. H. and Baron, S. (1971). *Infect. Immunity* **3**, 323.

59. Gober, L. L., Friedman-Kien, A. E., Havell, E. A. and Vilcek, J. Personal communication (1971).
60. Goorha, R. M. and Gifford, G. E. (1970). *Proc. Soc. exp. Biol. Med.* **134**, 1142.
61. Gresser, I. and Bourali, C. (1969). *Nature, Lond.* **223**, 844.
62. Gresser, I., Falcoff, R., Fontaine-Brouty-Boyé, D., Zajdela, F., Coppey, J. and Falcoff, E. (1967). *Proc. Soc. exp. Biol. Med.* **126**, 791.
63. Gresser, I., Bourali, C., Levy, J. P., Fontaine-Brouty-Boyé, D. and Thomas, M. T. (1969). *Proc. natn. Acad. Sci. U.S.A.*, **63**, 51.
64. Hamilton, L. D., Babcock, V. I. and Southam, C. M. (1969). *Proc. natn. Acad. Sci. U.S.A.*, **64**, 878.
65. Hanna, L., Merigan, T. C. and Jawetz, E. (1966). *Proc. Soc. exp. Biol. Med.* **122**, 147.
66. Harel, L. and Montagnier, L. (1971). *Nature New Biology, Lond.* **229**, 106.
67. Heberling, R. L. and Kalter, S. S. (1970). *Proc. Soc exp. Biol. Med.* **135**, 717.
68. Herman, R. and Baron, S. (1971). Personal communication.
69. Hill, D. A., Baron, S., Levy, H. B., Bellanti, J., Buckler, C. E., Cannellos, G., Carbone, P., Chanock, R. M., DeVita, V., Guggenheim, M. A., Homan, E., Kapikian, A. Z., Kirschstein, R. L., Mills, J., Perkins, J. C., Van Kirk, J. E. and Worthington, M. (1971). Perspectives in Virology VII. (M. Pollard, ed.), p. 197. Academic Press, New York and London.
70. Huppert, J., Hillova, J. and Gresland, L. (1969). *Nature, Lond.* **223**, 1015.
71. Jahiel, R. I., Vilcek, J., Nussenzweig, R. and Vanderberg, J. (1968). *Science, N.Y.* **161**, 802.
72. Jullien, P. and De Maeyer-Guignard, J. (1971). *Int J. Cancer* **7**, 468.
73. Kimball, P. C. and Duesberg, P. H. (1971). *J. Virol.* **7**, 697.
74. Kleinschmidt, W. J., Ellis, L. F., Van Frank, R. M. and Murphy, E. B. (1968). *Nature, Lond.* **220**, 167.
75. Kleinschmidt, W. J., Douthart, R. J. and Murphy, E. B. (1970). *Nature, Lond.* **228**, 27.
76. Lampson, G. P., Tytell, A. A., Field, A. K., Nemes, M. M. and Hilleman, M. R. (1967). *Proc. natn. Acad. Sci. U.S.A.*, **58**, 782.
77. Lampson, G. P., Tytell, A. A., Field, A. K., Nemes, M. M. and Hilleman, M. R. (1969). *Proc. Soc. exp. Biol. Med.* **132**, 212.
78. Lampson, G. P., Field, A. K., Tytell, A. A., Nemes, M. M. and Hilleman, M. R. (1970). *Proc. Soc. exp. Biol. Med.* **135**, 911.
79. Larson, V. M., Clark, W. R., Dagle, G. E. and Hilleman, M. R. (1969a). *Proc. Soc. exp. Biol. Med.* **132**, 602.
80. Larson, V. M., Clark, W. R. and Hilleman, M. R. (1969b). *Proc. Soc. exp. Biol. Med.* **131**, 1002.
81. Larson, V. M., Panteleakis, P. N. and Hilleman, M. R. (1970). *Proc. Soc. exp. Biol. Med.* **133**, 14.
82. Lemke, P. A. and Ness, T. M. (1970). *J. Virol.* **6**, 813.
83. Leonard, B. J., Eccleston, E. and Jones, D. (1969). *Nature, Lond.* **224**, 1023.
84. Levy, H. B. and Riley, F. (1970). *Proc. Soc. exp. Biol. Med.* **135**, 141.
85. Levy, H. B., Law, L. W. and Rabson, A. S. (1969). *Proc. natn. Acad. Sci. U.S.A.*, **62**, 357.
86. Lindh, H. F., Lindsay, H. L., Mayberry, B. R. and Forbes, M. (1969). *Proc. Soc. exp. Biol. Med.* **132**, 83.

87. Lindsay, H. L., Trown, P. W., Brandt, J. and Forbes, M. (1969). *Nature, Lond.* **223**, 717.
88. Lockart, R. Z., Jr., Bayliss, N. L., Toy, S. T. and Yin, F. H. (1968). *J. Virol.* **2**, 962.
89. Lomniczi, B. and Burke, D. C. (1970). *J. gen. Virol.* **8**, 55.
90. Margolis, S. A., Oie, H., Levy, H. B. and Baron, S. (1971). *Bact. Proc.* **195**.
91. Mathé, G., Amiel, J. L., Schwarzenberg, L., Schneider, M., Hayat, M., De Vassal, F., Jasmin, C., Rosenfeld, C., Sakouhi, M. and Choay, J. (1970). *Eur. J. Clin. Biol. Res.* **15**, 671.
92. Meier, H., Myers, D. D. and Huebner, R. J. (1970). *Life Sci.* **9**, 653.
93. Nemes, M. M., Tytell, A. A., Lampson, G. P., Field, A. K. and Hilleman, M. R. (1969a). *Proc. Soc. exp. Biol. Med.* **132**, 776.
94. Nemes, M. M., Tytell, A. A., Lampson, G. P., Field, A. K. and Hilleman, M. R. (1969b). *Proc. Soc. exp. Biol. Med.* **132**, 784.
94a. Nordlund, J. J., Wolff, S. M. and Levy, H. B. (1970). *Proc. Soc. exp. Biol. Med.* **133**, 439.
95. Oh, J. O. and O'Connor, G. R. (1971). *Infect. Immunity* **4**, 407.
96. Oh, J. O., Ostler, H. B. and Schachter, J. (1970). *Infect. Immunity* **1**, 566.
97. Ostler, H. B., Oh, J. O., Dawson, C. R. and Burt, W. L. (1970). *Nature, Lond.* **228**, 362.
98. Park, J. H. and Baron, S. (1968). *Science, N.Y.* **162**, 811.
99. Pieroni, R. E., Bundeally, A. E. and Levine, L. (1971). *J. Immun.* **106**, 1128.
100. Pitha, J. and Pitha, P. M. (1971). *Science, N.Y.* **172**, 1146.
101. Planterose, D. N., Birch, P. J., Pilch, D. J. F. and Sharpe, T. J. (1970). *Nature, Lond.* **227**, 504.
102. Postic, B. and Sather, G. E. (1970). *Ann. N.Y. Acad. Sci.* **173**, 606.
103. Regelson, W. and Munson, A. E. (1970). *Ann. N.Y. Acad. Sci.* **173**, 831.
104. Regelson, W., Munson, A. E., Wooles, W. R., Lawrence, W., Jr. and Levy, H. (1970). Colloques de l'Institut National de la Santé et de la Recherche Médicale, no. 6, "L'Interféron" p. 381.
105. Remington, J. S. and Merigan, T. C. (1968). *Science, N.Y.* **161**, 804.
106. Remington, J. S. and Merigan, T. C. (1970). *Nature, Lond.* **226**, 361.
107. Renis, H. E. (1970). *Appl. Microbiol.* **20**, 821.
108. Rhim, J. S. and Huebner, R. J. (1971). *Proc. Soc. exp. Biol. Med.* **136**, 524.
109. Rice, J. M., Turner, W., Chirigos, M. A. and Rice, N. R. (1970). *Appl. Microbiol.* **19**, 867.
110. Rosztoczy, I. and Mécs, I. (1970). *Acta virol., Prague* **14**, 398.
111. Sarma, P. S., Shiu, G., Neubauer, R. H., Baron, S. and Huebner, R. J. (1969). *Proc. natn. Acad. Sci. U.S.A.*, **62**, 1046.
111a. Stern, R. (1970). *Biochem. biophys. Res. Commun.* **41**, 608.
112. Stewart, W. E., II, Gosser, L. B. and Lockart, R. Z., Jr. (1971). *J. gen. Virol.* **13**, 35.
113. Stinebring, W. R. and Absher, M. (1970). Colloques de l'Institut National de la Santé et de la Recherche Médicale, no. 6, "L'Interferon" p. 63.
114. Tan, Y. H., Armstrong, J. A., Ke, Y. H. and Ho, M. (1970). *Proc. natn. Acad. Sci. U.S.A.*, **67**, 464.
115. Tan, Y. H., Armstrong, J. A. and Ho, M. (1971). *Virology* **44**, 503.
116. Tilles, J. G. (1970). *Proc. Soc. exp. Biol. Med.* **133**, 1334.
117. Turner, W., Chan, S. P. and Chirigos, M. A. (1970). *Proc. Soc. exp. Biol. Med.* **133**, 334.

118. Tytell, A. A., Lampson, G. P., Field, A. K. and Hilleman, M. R. (1967). *Proc. natn. Acad. Sci. U.S.A.*, **58**, 1719.
119. Tytell, A. A., Lampson, G. P., Field, A. K., Nemes, M. M. and Hilleman, M. R. (1970). *Proc. Soc. exp. Biol. Med.* **135**, 917.
120. Vandeputte, M., Datta, S. K., Billiau, A. and De Somer, P. (1970). *Eur. J. Cancer* **6**, 323.
121. Vilcek, J. (1970). *Ann. N.Y. Acad. Sci.* **173**, 390.
122. Vilcek, J. and Ng, M. H. (1971). *J. Virol.* **7**, 588.
123. Vilcek, J., Rossman, T. G. and Varacalli, F. (1969). *Nature, Lond.* **222**, 682.
124. Vilcek, J., Ng, M. H., Friedman-Kien, A. E. and Krawciw, T. (1968). *J. Virol.* **2**, 648.
125. Weinstein, M. J., Waitz, J. A. and Came, P. E. (1970). *Nature, Lond.* **226**, 170.
126. Weinstein, A. J., Gazdar, A. F., Sims, H. L. and Levy, H. B. (1971). *Nature New Biology, Lond.* **231**, 53.
127. Wheelock, E. F. and Larke, R. P. B. (1968). *Proc. Soc. exp. Biol. Med.* **127**, 230.
128. Winchurch, R. and Braun, W. (1969). *Nature, Lond.* **223**, 843.
129. Woodhour, A. F., Friedman, A., Tytell, A. A. and Hilleman, M. R. (1969). *Proc. Soc. exp. Biol. Med.* **131**, 809.
130. Worthington, M. and Baron, S. (1971). *Proc. Soc. exp. Biol. Med.* **136**, 323.
131. Young, P. A., Taylor, J. J., Yu, M. C. and Eyerman, E. (1970). *Nature, Lond.* **228**, 1191.
132. Youngner, J. S. and Hallum, J. V. (1968). *Virology* **35**, 177.
133. Youngner, J. S. and Hallum, J. V. (1969). *Virology* **37**, 473.
134. Zedeck, M. S., Marquardt, H., Sternberg, S. S., Fleisher, M. and Hamilton, L. D. (1970). *Proc. natn. Acad. Sci. U.S.A.*, **67**, 180.
135. Zeleznick, L. D. and Bhuyan, B. K. (1969). *Proc. Soc. exp. Biol. Med.* **130**, 126.

## DISCUSSION

**B. Zmudzka:** Do you think it is reasonable to interpret the activation of a complex like poly(rI) poly(rC) in terms of the model you proposed for thermal activation of poly r(A-U) in the interferon induction system?

**E. De Clercq:** I agree that the model proposed for thermal activation (modification of a looped, or "Christmas-tree" like, structure to a rod-like structure) can only be applied to alternating copolymers, and not to homopolymer pairs. It should, however, be pointed out that the extent of thermal activation is much greater with alternating copolymers than with homopolymer pairs: the shift in antiviral activity is five logs for alternating copolymers and only two for homopolymer pairs.

**C. J. Lucas:** What is the origin of the RNase sensitivity of poly(rI) . poly(rC)? Have you any idea as to the origin of the increased resistance to RNase of poly(rI) . poly(rC) after preincubation in tissue culture medium? And why culture medium for this purpose?

**E. De Clercq:** The RNase sensitivity of poly (rI) . poly (rC) resides in the poly (rC) portion, which is sensitive to pancreatic RNase. Even after complexing to poly (rI), poly (rC) is still sensitive to RNase, although less so. Poly (rI) is not sensitive to pancreatic RNase, but is susceptible to RNase $T_1$.

The mechanism of RNase resistance has been studied with poly r(A-U), an alternating copolymer, and is probably related to a change in its secondary structure. At 0°C poly r(A-U) would be present in the form of a branched structure with loops, in which the branching points are sensitive to pancreatic RNase. At 37°C poly r(A-U) would be in the form of a rod-like structure, which is resistant to pancreatic RNase.

Tissue culture medium is ideal for showing the thermal activation step. Thermal activation can also be demonstrated in a Tris-buffer medium containing divalent cations such as $Ca^{2+}$, $Mg^{2+}$.

**D. Shugar:** Can you comment on the findings of Colby and Chamberlin (*Proc. natn. Acad. Sci. U.S.A.* **63**, 160, 1969), who present reasonable evidence to the effect that lability to pancreatic RNase of synthetic polyribonucleotides is of little or no significance with regard to their interferon inducing activity?

**E. De Clercq:** Colby and Chamberlin did not find a correlation between antiviral activity and resistance to RNase with a variety of polyribonucleotides tested. However, there are different systems which have been used to increase the antiviral activity of polyribonucleotides. In these systems, there is a parallel increase in resistance to nuclease degradation. The systems employed are the following: (a) substitution of thiophosphate for phosphate, (b) preincubation of the polymers at 37°C, (c) complex formation with DEAE-dextran, (d) substitution of vinyl for ribophosphate.

Pertinent to this question are the observations of Stern, Friedman, Nordlund, etc., who have shown the presence of a nuclease in the serum of different animal species which specifically degrades double-stranded RNA. Stern (*Biochem. biophys. Res. Commun.* **41**, 608, 1970) partially purified the enzyme, and Friedman, Barth and Stern (*Nature, New Biology* **230**, 17, 1971) demonstrated that its nucleolytic activity was related to its inactivating effect on the antiviral activity of (poly rI) . (poly rC).

**W. Prusoff:** Since antiviral activity of synthetic polyribonucleotides is related to their molecular weight, is it not important to know the molecular weight of the 2'-O-methyl poly (rC) you referred to in relation to poly (rC) itself?

**E. De Clercq:** The poly (2'-O-MeC) mentioned in my talk, and which came from the laboratory of Shugar and Zmudzka, had an $S_{20,W}$ value exceeding eight, which corresponds well with the size of the poly (rC) samples used. Consequently, on a molecular weight basis, the activity of poly (rI) . poly (2'-O-MeC) is indeed much lower than that of poly (rI) . poly (rC).

Similarly poly (rI) . poly (2'-chloroC) has been found to be significantly less active than poly (rI) . poly (rC), notwithstanding that the molecular weights of the poly (rC) and the poly (2'chloroC) are comparable (personal communication from Black, Eckstein, Hobbs, Sternbach and Merigan, 1971).

# Studies with Temperature-sensitive Mutants of Adenovirus Type 5

N. M. WILKIE, S. USTACELEBI and J. F. WILLIAMS

*Medical Research Council, Virology Unit,*
*Institute of Virology, Glasgow, Scotland*

## INTRODUCTION

The adenovirus group consists of some 55 known serotypes, of which 31 derive from human hosts [22]. The virion is icosahedral [12] with a central nucleoprotein core and an external capsid consisting of 12 pentons at the vertices, and 240 hexons on the facets of the icosahedron [25]. Each penton is composed of two parts, the penton base and the fibre [25]. The DNA is a linear duplex with a molecular weight of 20-25 × $10^6$ [10], so that it has sufficient genetic information to code for some 20-50 proteins. Nine polypeptides have been resolved by acrylamide gel electrophoresis of purified type 2 virus particles by Maizel et al. [15], although only five polypeptides have been found in similar preparations from type 5 adenovirions by Russell et al. [20]. Thus approximately two thirds of the genome is available to code for non-structural proteins, most of which are presumably implicated in virus replication.

Temperature-sensitive (*ts*) mutants of bacteriophages [2, 3] have proven to be extremely useful in studying bacteriophage genetics and development [5], and a similar approach has begun to be applied to animal virology, [6]. The genetic functions of human adenoviruses have so far not been studied in this way, although cytocidal mutants of adenovirus type 12, some of which are host dependent, have been described by Takemori et al. [23]. Recently, we have isolated *ts* mutants of type 5 adenovirus from stocks mutagenized by nitrous acid or hydroxylamine *in vitro* and 5-bromodeoxyuridine *in vivo* [27]. There have also been recent reports of the isolation of *ts* mutants of adenovirus type 31 by Ito and Suzuki [14], avian adenovirus by Ishibashi [13], and type 5 human adenovirus by Ensinger and Ginsberg [4].

Temperature-sensitive mutations could, theoretically, affect all indispensable viral proteins, structural or non-structural. Thus *ts* mutants are potentially extremely valuable for investigating adenovirus-specified proteins, for determining their roles in virus replication, and in viral control of host cell macromolecular synthesis. This paper reports genetic studies with the type 5 *ts* mutants, provides data concerning their physiological characterization in terms of viral DNA synthesis, and discusses work concerning the proteins of these mutants. In addition, studies on transformation of rat embryo cells, and interferon induction in chick cells by these mutants will be discussed.

## GENETIC STUDIES: COMPLEMENTATION AND RECOMBINATION

The details of the genetic tests have been published by Williams and Ustacelebi [28, 29] and only an outline will be given here. For complementation, HeLa cells were infected with mutants, either singly at an input multiplicity of 10 p.f.u./cell, or doubly at an input of 5 p.f.u./cell for each mutant (total of 10 p.f.u./cell). The infected cells were incubated at the non-permissive temperature of 38° for 40 hours. Samples harvested at this time were then assayed on HeLa cells for infectivity at 31° using an improved plaque assay [26]. Complementation was measured by comparing the 38°C yields of virus in the single and double infections. At present, we consider complementation positive when the ratio of the yield of the double infection/yield of the higher of the two single infection, exceeds ten. In addition, yields at the permissive temperature of 31°C have also been determined in parallel with the 38°C yields.

The results of an experiment involving ten mutants are shown in Table 1. The values on the diagonal are the single infection yields, while those at the bottom of the table are the corresponding 31°C yields. Comparison of these values indicates that most of these mutants show very little leakiness at 38°C, and are well-suited for complementation analysis. According to the tenfold criterion mentioned above, all mutants complement each other in every combination, with the exception of the cross *ts* 2 x *ts* 14, so that the ten mutants must be assigned to nine complementation groups. The yield of this cross is only six-fold higher than that of the *ts* 14 single infection, but in repeated experiments with these two mutants the ratio was less than two, and we consider them to belong to the same complementation group. Certain crosses complement much more efficiently than others, and in some cases the efficiency is extremely high. For instance, in the case of the cross *ts* 2 and *ts* 4, there was a $1.1 \times 10^6$ fold increase in yield due to complementation, and this was as high as 40% of the single infection yield at 31°C. Other mutants have been tested for complementation and preliminary results indicate that we might have several other complementation groups.

For recombination tests, cells were singly or doubly infected as for complementation, and then incubated at 31°C for 5 days. The samples collected at this time were assayed at 31°C for total infectivity and at 38°C to measure production of wild type progeny. The frequency of wild-type recombinants in the yields is the ratio, (infectivity measured at 38°C)/(infectivity measured at

**Table 1.** Complementation with *ts* mutants of adenovirus type 5: Cells were infected either with pairs of mutants each at input multiplicities of 5 p.f.u./cell, or with single mutants at input multiplicities of 10 p.f.u./cell. Infected cultures were incubated for 40 hr at 38°C, after which time virus yields were assayed at 31°C. In this table the yields are expressed as p.f.u./ml. The values in the bottom line are the 31°C yields of single infections assayed at 31°C.

|           | 5 ts 1 | 5 ts 2 | 5 ts 3 | 5 ts 4 | 5 ts 13 | 5 ts 14 | 5 ts 17 | 5 ts 18 | 5 ts 19 | 5 ts 22 |
|-----------|--------|--------|--------|--------|---------|---------|---------|---------|---------|---------|
| 5 ts 1    | $<10^2$ | $5.6 \times 10^7$ | $2.1 \times 10^4$ | $2.6 \times 10^4$ | $1.8 \times 10^8$ | $1.7 \times 10^7$ | $3.0 \times 10^4$ | $3.2 \times 10^7$ | $5.4 \times 10^6$ | $2.8 \times 10^8$ |
| 5 ts 2    | —      | $<10^2$ | $5.0 \times 10^7$ | $1.1 \times 10^8$ | $2.2 \times 10^8$ | $1.2 \times 10^3$ | $9.0 \times 10^7$ | $1.0 \times 10^4$ | $8.2 \times 10^7$ | $1.7 \times 10^8$ |
| 5 ts 3    | —      | —      | $<10^2$ | $2.2 \times 10^4$ | $1.2 \times 10^8$ | $1.4 \times 10^7$ | $2.3 \times 10^4$ | $1.9 \times 10^7$ | $2.1 \times 10^6$ | $2.0 \times 10^8$ |
| 5 ts 4    | —      | —      | —      | $<10^2$ | $1.6 \times 10^8$ | $4.1 \times 10^7$ | $5.8 \times 10^4$ | $2.3 \times 10^7$ | $7.6 \times 10^6$ | $3.0 \times 10^8$ |
| 5 ts 13   | —      | —      | —      | —      | $2.8 \times 10^5$ | $7.5 \times 10^6$ | $6.0 \times 10^6$ | $4.4 \times 10^6$ | $3.1 \times 10^6$ | $5.7 \times 10^7$ |
| 5 ts 14   | —      | —      | —      | —      | —       | $2.0 \times 10^2$ | $1.2 \times 10^7$ | $1.2 \times 10^4$ | $1.4 \times 10^7$ | $1.0 \times 10^8$ |
| 5 ts 17   | —      | —      | —      | —      | —       | —       | $4.5 \times 10^2$ | $4.2 \times 10^6$ | $1.0 \times 10^6$ | $1.4 \times 10^8$ |
| 5 ts 18   | —      | —      | —      | —      | —       | —       | —       | $<10^2$ | $3.7 \times 10^7$ | $1.9 \times 10^8$ |
| 5 ts 19   | —      | —      | —      | —      | —       | —       | —       | —       | $1.9 \times 10^4$ | $1.6 \times 10^8$ |
| 5 ts 22   | —      | —      | —      | —      | —       | —       | —       | —       | —       | $1.5 \times 10^5$ |
| at 31°C   | $1.3 \times 10^8$ | $2.0 \times 10^8$ | $1.0 \times 10^8$ | $1.8 \times 10^8$ | $1.5 \times 10^8$ | $1.3 \times 10^8$ | $9.0 \times 10^7$ | $5.0 \times 10^7$ | $3.5 \times 10^7$ | $1.5 \times 10^8$ |

31°C), expressed as a percentage. Results of a recombination experiment involving four *ts* mutants are shown in Table 2. There is no evidence of back-mutation in these mutants, with the percentage of wild-type virus in the 31°C yields of single infections being $< 3.7 \times 10^{-4}$. The actual proportion is probably lower but, due to a toxic effect of adenovirus *ts* mutants at 38°C at high multiplicity, only the upper limits could be estimated. In order to establish that recombination was really occurring, progeny tests were carried out. For this, wild-type plaques were picked from assay plates at 38°C, and these were reassayed at 38°C and 31°C in order to determine if they bred true as wild-type

**Table 2.** Recombination with *ts* mutants of adenovirus type 5: Cells were infected as for complementation at total input multiplicities of 10 p.f.u./cell, and were incubated for 5 days at 31°C. Virus yields were assayed at 38°C and 31°C. The values in the table represent the percentages of wild-type recombinants in the yields. Figures in brackets are the ratios—number of progeny plates behaving as wild-type/number of plaques tested.

|        | 5 *ts* 1 | 5 *ts* 2 | 5 *ts* 3 | 5 *ts* 4 |
|--------|----------|----------|----------|----------|
| 5 *ts* 1 | $< 1.4 \times 10^{-4}$ | 7.60 (15/15) | 1.08 (16/16) | *0.46 (12/13) |
| 5 *ts* 2 | – | $< 1.8 \times 10^{-4}$ | 5.43 (14/14) | *4.20 (11/13) |
| 5 *ts* 3 | – | – | $< 3.7 \times 10^{-4}$ | 1.10 (13/13) |
| 5 *ts* 4 | – | – | – | $< 2.1 \times 10^{-4}$ |

* corrected value allowing for progeny test result.

virus. An infectivity ratio (38°C/31°C) of $1.0 \pm 0.5$ was considered to show that the progeny were wild-type, while a ratio of $< 0.5$ was taken to indicate that the progeny were not normal recombinants. Results of the progeny tests on these crosses are given in brackets under the frequencies given in Table 2. As can be seen, tested plaques scored as wild-type with the exception of two from cross *ts* 2 x *ts* 4 and one from cross *ts* 1 x *ts* 4 which scored as *ts*. We have observed that the absolute recombination frequencies vary from experiment to experiment [29] while the relative frequencies remain fairly constant. This problem is now being investigated further. On the basis of the preliminary data the four mutants can be arranged in an additive sequence *ts* 1-*ts* 4-*ts* 3-*ts* 2.

## PHYSIOLOGY OF THE MUTANTS

Experiments were carried out to ascertain if the *ts* mutants synthesized viral DNA in infected cells at the non-permissive temperature. Monolayers of HeLa cells were infected at a multiplicity of infection of 10-20 p.f.u per cell at 38°C.

The cells were exposed to $^3$H-thymidine at various times post infection, and total intracellular DNA isolated following lysis with SDS and digestion with pronase as described by Pina and Green [19]. $^{32}$P-labelled DNA isolated from purified adenovirus was added as a marker, and the DNAs separated by banding on CsCl density gradients. Mock-infected cells labelled from 0-9 or 16-24 hours post infection showed only a single peak of $^3$H-DNA which banded at a lower density than the $^{32}$P-labelled virus DNA (Fig. 1a). This material actually had a density of 1.6981 which is characteristic of human DNA (Fig. 2a). A similar result was obtained with DNA from cells labelled from 0-9 hours after infection with wild type virus. However, when thymidine was added from 16-24 hours

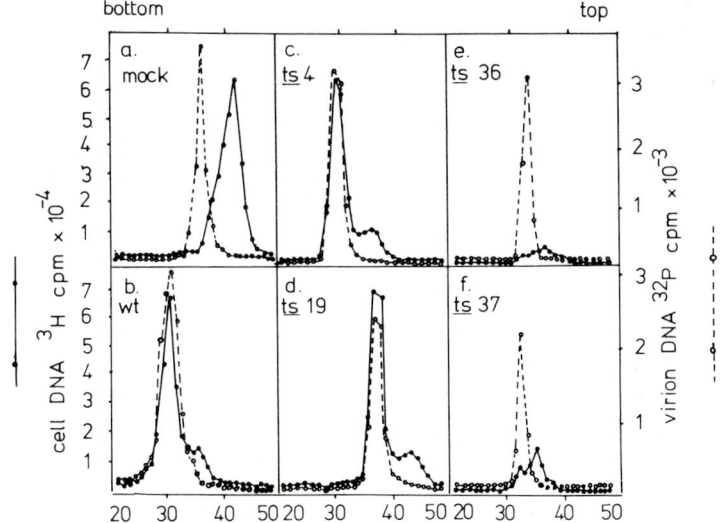

**Figure 1.** CsCl buoyant density separation of DNA extracted from HeLa cells exposed to $^3$H thymidine from 16-24 hours after infection with adenovirus type 5. $^{32}$PDNA from adenovirus 5 added as a marker.

post infection, it was observed that host DNA synthesis had been depressed and a new peak of $^3$H-DNA was obtained which coincided with the $^{32}$P marker (Fig. 1b). The same result was observed with *ts* 4 (Fig. 1c) and *ts* 19 (Fig. 1d) and in fact with all of the mutants listed in Table 1.

Thirteen other *ts* mutants which had not been tested for complementation were also analysed for the ability to synthesize viral DNA at the non-permissive temperature, and most gave the wild type pattern. However, with the mutants *ts* 36 and *ts* 37, practically no viral DNA synthesis was observed, although host DNA synthesis was depressed almost as much as by wild type (Fig. 1e, 1f).

These results were confirmed by banding unlabelled DNA, isolated from infected cells, in the analytical ultracentrifuge. Herpes simplex virus DNA ($\rho$ = 1.7254) was always included to act as a buoyant density marker, and the density (and consequently the G + C content) of each peak of DNA was calculated as described by Schildkraut *et al.* [21]. HeLa cell DNA banded at a density of 1.6981, but DNA extracted 24 hours after infection with wild type adenovirus 5 contained an additional peak of density 1.7137. This corresponds to a G + C content of 54.8%, which is characteristic of the DNA of non-oncogenic adenoviruses [18]. This band of DNA was observed with all of the mutants tested except *ts* 36 and *ts* 37, (Figs. 2d, 2e). In all of these experiments, no band of viral DNA could be detected up to nine hours after infection, confirming that the DNA from the input virus did not show up in the assay. It was concluded that *ts* 36 and *ts* 37 were unable to replicate viral DNA in HeLa cells at 38°C.

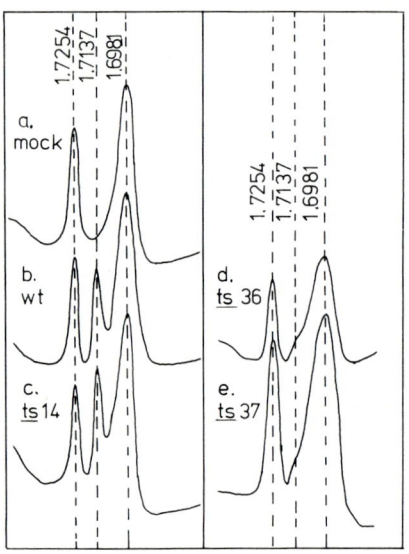

**Figure 2.** CsCl buoyant density separation in the analytical ultracentrifuge of unlabelled DNA extracted from HeLa cells 24 hours after infection with adenovirus type 5. Herpes simplex virus type I DNA added as a buoyant density marker.

Mutants *ts* 36 and *ts* 37 were then tested for their ability to complement each other and some of the other mutants, and the data are presented in Table 3. These two mutants complemented with the other four mutants *ts* 1, 2, 18, and 19, but not with each other. We conclude that *ts* 36 and *ts* 37 are mutations in the same gene. Both of these mutants show a high degree of leakiness, and this may have contributed to the apparently low increase in mixed-infection yields compared with single infection yields of those two with the others. We are now attempting to isolate less leaky clones of *ts* 36 and *ts* 37 for use in complementation and physiological studies.

It has been reported that the major capsid antigens are synthesized in adenovirus type 5 infected cells some time after the onset of viral DNA synthesis

[16] and that another viral antigen, the P antigen, could be detected prior to viral DNA synthesis. Protein synthesis at the restrictive temperature with a number of our type 5 *ts* mutants is being examined by Russell at Mill Hill, and he has kindly allowed us to discuss some of his results. He has been comparing the synthesis of viral proteins in cells infected with wild-type and *ts* mutants at the non-permissive temperature, using both immunological techniques and analytical procedures for separation of polypeptides on polyacrylamide gels. He finds that most of the mutants tested to date synthesize normal amounts of the main viral antigens. With *ts* 37 however, large amounts of P antigen appear, but only very low amounts of hexon, penton base and fibre. The small amounts of these antigens could result from the leakiness of this mutant. This finding fits in very well with the finding that *ts* 37 fails to make DNA, and agrees with the previous observation that these antigens appear only after viral DNA synthesis has occurred. It seems likely that the mRNA for the P antigen is transcribed from the parental virus DNA.

**Table 3.** Complementation between DNA-positive and DNA-negative *ts* mutants.

|  | 5 *ts* 1 | 5 *ts* 2 | 5 *ts* 18 | 5 *ts* 19 | 5 *ts* 36 | 5 *ts* 37 |
|---|---|---|---|---|---|---|
| 5 *ts* 1 | $< 10^3$ | $1.6 \times 10^7$ | $9.0 \times 10^6$ | $9.2 \times 10^7$ | $7.5 \times 10^8$ | $7.6 \times 10^8$ |
| 5 *ts* 2 | – | $1.5 \times 10^3$ | – | $1.8 \times 10^8$ | $5.0 \times 10^8$ | $6.0 \times 10^8$ |
| 5 *ts* 18 | – | – | $4.2 \times 10^4$ | $3.4 \times 10^8$ | $8.0 \times 10^8$ | $6.7 \times 10^8$ |
| 5 *ts* 19 | – | – | – | $3.0 \times 10^6$ | $7.5 \times 10^8$ | $1.0 \times 10^9$ |
| 5 *ts* 36 | – | – | – | – | $4.6 \times 10^7$ | $8.6 \times 10^7$ |
| 5 *ts* 37 | – | – | – | – | – | $1.02 \times 10^8$ |
| at 31°C | $3.2 \times 10^9$ | $3.0 \times 10^9$ | $1.4 \times 10^9$ | $1.4 \times 10^9$ | $1.6 \times 10^9$ | $1.4 \times 10^9$ |

Russell has observed that *ts* 13 makes normal amounts of hexon and penton base at the restrictive temperature, but no detectable fibre antigen or fibre polypeptide. It remains to be determined whether this is due to a block in transcription or in translation. This mutant is of obvious practical importance, since it allows penton base to be isolated free from fibre, and in addition it should be useful in testing whether fibre antigen is responsible for inhibiting host cell macromolecular synthesis [9]. We have already observed that *ts* 13, while failing to make fibre, does inhibit cell DNA synthesis. Several of the *ts* mutants appear to involve transport functions. Normal amounts of virion proteins are synthesized, but their transport from the cytoplasm to the nucleus appears to be blocked.

Extension of our genetic and biochemical studies should allow us to obtain a clearer picture of the molecular events involved in virus replication and maturation.

## TRANSFORMATION BY ADENOVIRUS TYPE 5

The group C adenoviruses (types 1, 2 and 5) transform rodent cells with a low frequency and at low efficiency [7, 17]. We have confirmed that adenovirus type 5 transforms rat embryo cells [28] and have established conditions which give fairly reproducible results, with transformation frequencies of around $5 \times 10^{-6}$ to $10^{-5}$ (i.e. the ratio of number of transformed foci/number of infected cells). The morphological appearance of a transformed focus, induced by adenovirus 5, is illustrated in Fig. 3. Individual foci, isolated from infected cultures, have been grown for up to twenty consecutive weekly passages, and tested for malignancy by both subcutaneous and intraperitoneal inoculation into young adult rats. To date, none of the test focal lines have given rise to tumours, and further tests are in progress to determine whether there is malignant transformation by type 5 adenovirus. Neither type 2 [7] nor type 1 [17] transformed cells appear to be capable of growing to form tumours in rats.

We are using the *ts* mutants of adenovirus type 5 to test the hypothesis that transformation depends upon continued expression of viral genetic material in the transformed cell. There is evidence that only a few of the 20-50 adenovirus genes operate in transformed cells [8] and, provided these genes also specify functions essential to virus replication, *ts* mutants should allow us to identify them. It should be possible to determine the role of these gene products in initiating and maintaining the altered behaviour of the transformed cell. To date, we have tested ten mutants for their ability to transform cells at the non-permissive temperature of 38°C, but it appears that all ten are able to transform cells with a frequency similar to that of wild-type virus. Other mutants, including the two DNA negative ones, are now being tested for their ability to transform rat cells at the non-permissive temperature.

## INTERFERON INDUCTION BY ADENOVIRUS TYPE 5

Several different types of human adenoviruses induce interferon in chick embryo cells [1, 11]. The adenovirus function(s) responsible for inducing interferon have not been identified, and the *ts* mutants of type 5 adenovirus provide us with a possible system for investigating this problem. Ten *ts* mutants have been tested for their ability to induce interferon in chick embryo at both permissive (31°C) and non-permissive (38°C) temperatures [24]. The results of two experiments are shown in Table 4, clearly *ts* 18 and *ts* 19 fail to induce interferon at 38°C. We have found efficient complementation between these two mutants (see Table 1), which suggests that these mutations are located in different genes which probably control different viral functions. Tests to determine whether *ts* 18 and *ts* 19 complement each other in chick embryo cells have been negative in that no interferon is induced in doubly infected cells. This

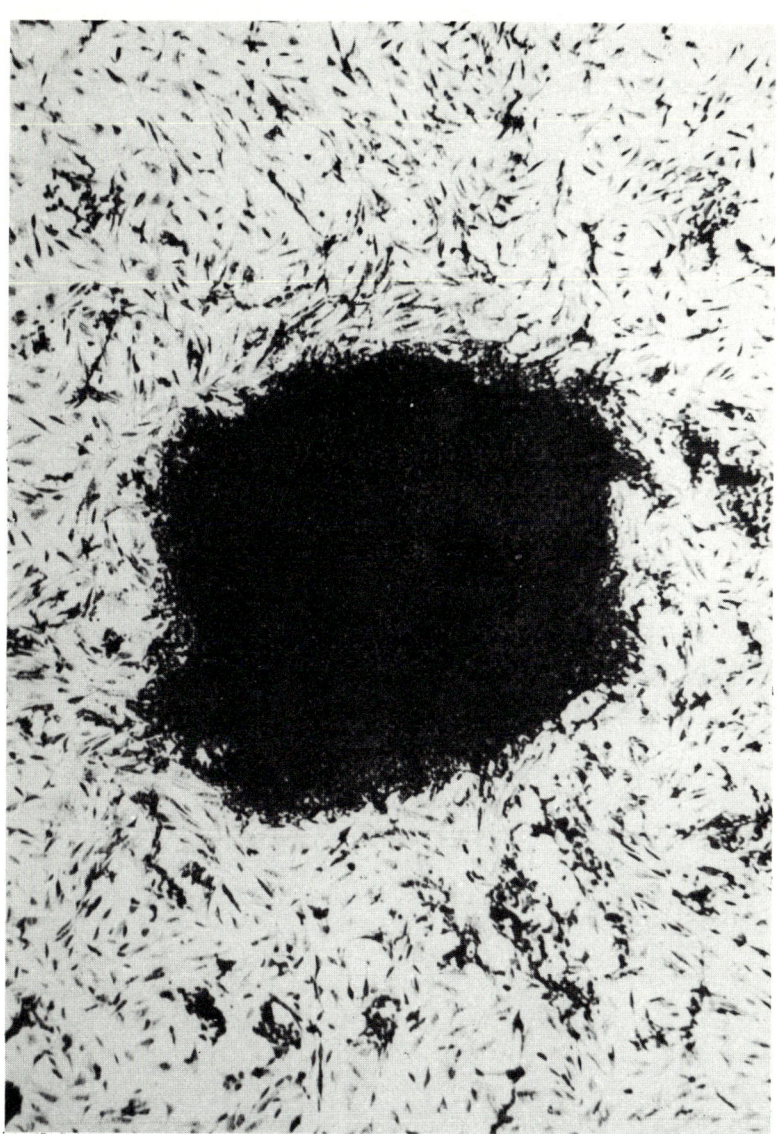

**Figure 3.** Colony of rat embryo cells transformed by adenovirus type 5. Culture stained with Giemsa five weeks after infection.

problem is being investigated further by temperature shift and biochemical experiments. The combined genetic-biochemical approach might lead to identification of the viral function(s) involved in interferon induction.

**Table 4.** Induction of interferon at $31°C$ and $38°C$ by wild type and *ts* mutants of adenovirus type 5.

| Virus | Interferon titre* | | | |
|---|---|---|---|---|
| | Experiment 1 | | Experiment 2 | |
| | $31°C$ | $38°C$ | $31°C$ | $38°C$ |
| 5 wild-type | 128 | 128 | 256 | 128 |
| 5 *ts* 1 | 256 | 128 | 256 | 128 |
| 5 *ts* 2 | 256 | 32 | 128 | 128 |
| 5 *ts* 3 | 256 | 128 | 128 | 64 |
| 5 *ts* 4 | 128 | 64 | 256 | 64 |
| 5 *ts* 13 | 256 | 128 | 128 | 64 |
| 5 *ts* 14 | 256 | 128 | 128 | 64 |
| 5 *ts* 17 | 256 | 128 | 128 | 64 |
| 5 *ts* 18 | 128 | $<4$ | 64 | $<4$ |
| 5 *ts* 19 | 64 | $<4$ | 64 | $<4$ |
| 5 *ts* 20 | 128 | 128 | 128 | 64 |

* PDD 50

## CONCLUDING REMARKS

In the preliminary complementation tests, the ten mutants fell into nine complementation groups. This is not entirely unexpected since the adenovirus genome has sufficient information to code for some 20-50 proteins of average size. The result is compatible with the view that mutations in most genes can be detected with equal facility and it seems unlikely that a highly mutable cistron is present. The data so far are not inconsistent with the view of Fujinaga and Green [8] that only a small part of the genome is involved in early functions; only one group defective in DNA replication has been found, and most mutants appear to produce the capsid antigens. It will be of great interest to determine, eventually, how many other virus genes are essential to DNA replication.

The mutants should be extremely useful in locating and identifying the physiological functions of the virus genes involved in adenovirus replication, and in elucidating virus induced events in cells resulting from virus-cell interactions. Of particular interest in this latter respect will be the sets of gene functions involved in the transforming capacity and the interferon-inducing capacity of adenovirus, and it is very encouraging that two *ts* mutants defective in interferon induction at non-permissive temperature have already been isolated.

## SUMMARY

Temperature sensitive mutants of adenovirus type 5 have been isolated from virus stocks mutagenized with nitrous acid, hydroxylamine or 5-bromodeoxyuridine. Ten of the mutants were found to fall into nine complementation groups. A preliminary analysis of recombination has been made with four of the mutants at the permissive temperature, wild type recombinants formed in all mixed infections. Two mutants, ts 36 and ts 37 have been found which fail to complement each other, and which are defective in viral DNA synthesis in HeLa cells at the non-permissive temperature of 38°C. The ts mutants are being used to investigate the viral gene functions involved in the transformation of rat embryo cells by adenovirus type 5. All mutants tested to date transform rat cells as efficiently at the non-permissive temperature as does wild type virus. Finally, the ts mutants are being used to investigate the adenovirus function(s) responsible for inducing interferon in chick embryo cells. Two mutants, ts 18 and ts 19, which complement each other in HeLa cells, do not induce interferon at the non-permissive temperature.

## ACKNOWLEDGEMENTS

We thank Professor J. H. Subak-Sharpe for the interest he has shown in this work. We also wish to thank Mrs. Sylvia McDonald and Mr. Alan Revill for excellent technical assistance. S. Ustacelebi is in receipt of a grant from the Turkish Ministry of Education.

## REFERENCES

1. Béládi, I. and Pusztai, R. (1967). *Z. Naturf.* **226**, 165.
2. Campbell, A. (1961). *Virology* **14**, 22.
3. Edgar, R. S. and Lielausis, I. (1964). *Genetics* **49**, 649.
4. Ensinger, M. J. and Ginsberg, H. S. (1971). *Bact. Proc.* **331**, 223.
5. Epstein, R. H., Bolle, A., Steinberg, C. M., Kellenberger, E., Boyde la Tour, E., Chevalley, E., Edgar, R. S., Susman, M., Denhardt, G. T. and Lielausis, A. (1963). *Cold Spring Harb. Symp. quant. Biol.* **28**, 375.
6. Fenner, F. (1969). *Curr. Topics Microbiol. Immunol.* **48**, 1.
7. Freeman, A. E., Black, P. H., Vanderpool, E. A., Henry, P. H., Austin, J. B. and Anebner, R. J. (1967). *Proc. natn. Acad. Sci. U.S.A.* **58**, 1205.
8. Fujinaga, K. and Green, M. (1970). *Proc. natn. Acad. Sci. U.S.A.* **65**, 1375.
9. Ginsberg, H. S., Bello, L. J. and Levine, A. J. (1967). *In* "The Molecular Biology of Viruses" (Colter, S. S. and Paranchych, W., eds), pp. 347, Academic Press, New York and London.
10. Green, M., Pina, M., Kimes, R., Wensink, P. C., MacHattie, L. A. and Thomas, C. A. Jun. (1967). *Proc. natn. Acad. Sci. U.S.A.* **57**, 1302.
11. Ho, M. and Kohler, K. (1967). *Arch. Ges. Virusforsch.* **22**, 69.
12. Horne, R. W. Brenner, S., Waterson, A. P. and Wildy, P. (1959). *J. molec Biol.* **1**, 84.
13. Ishibashi, M. (1970) *Biken's J.* **13**, 59.

14. Ito, M. and Suzuki, E. (1970). *J. gen. Virol.* **9**, 243.
15. Maizel, J. V. Jun., White, D. O. and Schariff, M. D. (1968). *Virology* **36**, 115.
16. Mantyjarvi, R. and Russell, W. C. (1969). *J. gen. Virol.* **5**, 339.
17. McAllister, R. M., Nicholson, M. D., Lewis, A. M. Jun., MacPherson, I. and Heubner, R. J. (1969). *J. gen. Virol.* **4**, 115.
18. Pina, M. and Green, M. (1965). *Proc. natn. Acad. Sci., U.S.A.* **54**, 547.
19. Pina, M. and Green, M. (1969). *Virology* **38**, 573.
20. Russell, W. C., McIntosh, K. and Skehel, J. J. (1971). *J. gen. Virol.* **11**, 35.
21. Schildkraut, C. L., Marmur, J. and Doty, P. (1962). *J. molec. Biol.* **4**, 430.
22. Schlesinger, R. W. (1969). *Adv. Virus Res.* **14**, 1.
23. Takemori, M. Riggs, J. L. and Aldrich, C. (1968). *Virology* **38**, 8.
24. Ustacelebi, S. and Williams, J. F. (1971). *Nature, Lond.* **235**, 52.
25. Valentine, R. C. and Pereira, H. G. (1965). *J. molec. Biol.* **13**, 13.
26. Williams, J. F. (1970). *J. gen. Virol.* **9**, 251.
27. Williams, J. F., Gharpure, M., Ustacelebi, S. and McDonald (1971). *J. gen. Virol.* **11**, 95.
28. Williams, J. F. and Ustacelebi, S. (1971). Ciba Foundation Symposium "The Strategy of the Viral Genome." (Wolstenholme and O'Connor, eds), pp. 275, Churchill Livingstone, London.
29. Williams, J. F. and Ustacelebi, S. (1971). *J. gen. Virol.* **13**, 345.

# Specific Ribosomes, Components of an Oncogenic RNA Virus

JOSEF ŘÍMAN, JAN KORB and ALENA MICHLOVÁ

*Laboratory for Biochemical Investigation of Cancer,
Institute of Organic Chemistry and Biochemistry,
Czechoslovak Academy of Sciences, Prague*

## INTRODUCTION*

It has been shown that some "non-viral" RNA species may be present in the virions of avian myeloblastosis (AMV) [1, 2, 3], and the evidence for this is particularly good in the case of AMV-tRNAs, which exhibit some distinct differences when compared with tRNAs from the virus transformed or healthy chicken cells [4, 5, 6]. Another "non-viral" RNA species, reported to be present in AMV, resembles according to its sedimentation pattern the cellular rRNAs [2].

The question of the virus structure localization, cell organelle origin and the functional destiny of these "non-viral" RNAs led us to search for ribosomes in AMV.

Here we present a description of the experimental conditions which enabled us to detect specific ribosomal structures as a constant component of AMV.

## MATERIALS AND METHODS

### Labelling and Isolation of the Virus and its Nucleic Acids

Avian leukemic myeloblasts, cultivated as described earlier [7], void of any mycoplasma [7], incubated in 20 ml cell suspension ($10 \times 10^7$ cell/ml), were pulse-labelled for 60 min with $^3$H-thymidine, 250 µCi/ml (thymidine-methyl-T, Calatomic, specific activity 15 Ci/mmole) in the presence of deoxycytidine, 20 µg/ml, and cytidine, 20 µg/ml (both Calbiochem, A grade). Then the medium was replaced by a medium containing $^{14}$C-cytidine, 3.8 µCi/ml (cytidine-$^{14}$C (U), R.C. Amersham, specific activity, 211 µCi/mmole) and the cells were pulse-labelled for another 60 min. After the second pulse-labelling the medium was replaced by an unlabelled one which was changed four times at 90 min

---

\* The following abbreviations are used in this text: AMT buffer: 0.1 molar NH$_4$Cl, 0.01 molar MgCl$_2$, 0.01 molar Tris-HCl; Thd, dCd, Cd, Ud: thymidine, deoxycytidine, cytidine, uridine; DOC, sodium deoxycholate; CR: cytoplasmic ribosomes; VR: "virus ribosomes".

intervals. The virus was collected and purified according to Erickson [8] from the cell supernatants derived from each unlabelled medium exchange. The nucleic acids were then isolated from the individual viral pellets according to Bader and Steck [9] in the presence of unlabelled myeloblastic RNA [10] of a known sedimentation profile. This RNA served on the one hand as an absorbance marker, and on the other as an internal marker indicating the integrity of the extracted virus RNAs [10]. Extracted nucleic acids, twice precipitated with ethanol, were collected by centrifugation at 20,000 x $g$, 30 min and + 4°C and dissolved in a buffer containing 0.1 molar NaCl, 0.005 molar Tris-HCl (pH 8.5), and 0.001 molar EDTA; they were further analysed in glycerol density gradients supplemented with the same buffer. The sampled fractions, after addition of human serum albumine (Spofa) (500 $\mu$g/fraction) and of TCA to a final concentration of 5%, were filtered through millipore filters. The TCA insoluble radioactivity remaining on the filters was determined [7] by scintillation spectrophotometry (Packard, model 3375), using BBOT-Packard scintillation fluid.

**Labelling and Isolation of Virus Ribosomes**

*Virus labelling with $^{32}P$*: To achieve a partial cell synchronization before the pulse labelling, myeloblasts cell suspension (50 ml, 4.5 x $10^7$ cells/ml) was incubated in a complete medium for 16 hours in the presence of hydroxyures [11] (Sigma), 0.0025 molar. Subsequently the cells collected by centrifugation were suspended in 20 ml (1.2 x $10^\circ$ cells/ml) of a phosphate-free medium, supplemented with Calbiochem A grade deoxyadenosine, 28 mg; deoxyguanosine, 10.6 mg; deoxycytidine, 0.45 mg; thymidine, 0.86 mg, and with 5 mCi of $^{32}P$ (carrier free, orthophosphate, Amersham, specific activity 50 Ci/mg P) and incubated under these conditions for 2 hr. Then the medium was replaced by a complete and unlabelled one which was changed twice at 5 hr intervals and once more at a 7 hr interval. Cell supernatant derived from each medium exchange was clarified by centrifugation at 3000 x $g$, 30 min, frozen and stored at –78°C until use, but for not more than seven days.

*Virus labelling with $^3H$-uridine*: Myeloblasts, pre-treated with hydroxyurea as above, were suspended in 20 ml (9.5 x $10^7$ cells/ml) of a complete medium supplemented with all the deoxynucleosides [11] as described above and with $^3$H-uridine, 100 $\mu$Ci/ml (Ud-5-$^3$H, UVVR-Praha, specific activity 24.5 Ci/mmole). The pulse labelling was accomplished in 4 hr and the medium was replaced by an unlabelled one which was changed three times at 5 hr intervals. Cell supernatant derived from each unlabelled medium exchange was further prepared as in the case of $^{32}P$-virus labelling.

*Virus collection and isolation*: The frozen 20 ml samples of the supernatants containing the labelled virus were rapidly thawed and clarified by centrifugation at 5000 x $g$, 30 min, +4°C, and mixed with 5 ml of similarly prepared supernatant of heparinized leukaemic chicken blood plasma containing 6.5 × $10^{11}$ particles per ml. Clarification at 3000 × $g$ was repeated once more and the supernatant of the joint samples was layered on the top of the following discontinuous sucrose density gradient: 2 ml of 65; 2 ml of 52; 1 ml of 20; and 2 ml of 15% sucrose. In order to reduce possible virus contamination by exogenous ribosomes, the sucrose at all densities was supplemented with 0.1 molar NaCl; EDTA, 0.001 molar; and Tris-HCl, 0.05 molar, pH 7.5 [12]. After 120 min centrifugation at 25,000 rev/min and +4°C in a Spinco rotor SW 25.1, the virus pellet situated on he top of the 52% sucrose layer was carefully collected (15-20 drops), diluted to a final volume of 12 ml with a solution containing NaCl, 0.1 molar, Tris-HCl, 0.05 molar, pH 7.4 and 8.7% sucrose. This virus suspension was layered on the top of 20% sucrose (4 ml) supplemented with 1/2 AMT [13] and centrifuged at 45,000 × $g$, 90 min.

*VR isolation*: Virus pellet was suspended at +4°C in 0.95 ml of 1/4 AMT supplemented with spermidine [14] (spermidine phosphate trihydrate, C grade; Calbiochem) to a final concentration of 0.0001 molar. After addition of 0.05 ml of 10% sodium deoxycholate (DOC), and mixing with he aid of a Pasteur pipette, the suspension was centrifuged 30 min at 20,000 × $g$ and +4°C. Virus supernatant was carefully removed and stored at 0°C. The resulting sediment was treated once more in the same way. Both 20,000 × $g$ supernatants (about 1.8 ml) were combined and further analysed by sucrose density gradient centrifugation as described above. The radioactivity was determined in acid insoluble material of the sampled fractions as described on p. 100.

## Co-sedimentation Characteristics of Virus and Cytoplasmic Ribosomes

*Virus labelling*: The labelled VRs were isolated from the 5 hr old virions produced by cells constantly exposed 20 and 25 hr to the radioisotope. In the case of labelling with $^3$H-uridine, cells pretreated with hydroxyurea (see p. 100) were suspended in 20 ml (9 × $10^7$ cells/ml) of the complete medium, supplemented with the four deoxynucleosides and with $^3$H-uridine, 125 $\mu$Ci/ml (specific activity 24.5 Ci/mmole) and incubated 4 hr. Then the medium was replaced by a labelled medium without deoxynucleosides containing the same amount of radioisotope. This medium was changed at 5 hr intervals three times and the virus collected and purified from the combined cell supernatants derived from the second and third medium exchange. In the case of labelling with $^{32}$P, the hydroxyurea-treated cells were suspended in 25 ml (8.5 × $10^7$ cells/ml) of a phosphate-free medium supplemented with deoxynucleosides [11] and with $^{32}$P, 160 $\mu$Ci/ml (orthophosphate, carrier free, Isocommerz, Germany). In this medium the cells were incubated 120 min and the medium replaced by a new labelled (160 $\mu$Ci/ml of $^{32}$P) phosphate-free medium without deoxynucleosides. This medium was changed at 5 hr intervals three times and the virus collected and purified from the combined supernatants. Isolation and analysis of

$^3$H-uridine or $^{32}$P-labelled VRs was accomplished exactly as described on p. 100. In the case of $^{32}$P-labelled ribosomes, each of the sampled fractions was divided into two equal aliquots. One of these was used for absorbance determination, accomplished with the Zeiss spectrophotometer VSU-2P, and for radioactivity determination. The second sample, corresponding to the peak fraction of VR monomers, was used for a dissociation experiment, which was performed by adding distilled water so that the MgCl$_2$ concentration was 0.0005 molar [14]. Then the sample, cooled in ice, was homogenized and, after a clarifying spin and addition of 10 A$_{260}$ units of CR-monomer, submitted once more to sucrose density gradient centrifugation.

*Isolation of CRs*: To achieve conditions comparable to those used in the case of VRs, CRs of unlabelled chicken liver and spleen cells and of the labelled myeloblasts were isolated from the 20,000 x *g* supernatant of cells disrupted in a 1/4 AMT-sucrose solution according to Küntzel and Noll [13]. In this case the homogenizing medium was supplemented with spermidine at 0.0001 molar concentration. After addition of DOC to a final concentration of 0.5%, the 20,000 x *g* cell supernatant was centrifuged once more at 20,000 x *g* for 15 min at +4°C. The clarified supernatant containing the ribosomes was layered over 15% sucrose, supplemented with AMT and centrifuged 7 hr at 130,000 x *g*. Then the translucent ribosomal pellets were suspended by homogenization in a solution containing 1/4 AMT, spermidine, 0.0001 molar and polyvinylpyrrolidone, 2%. After a clarifying spin the CR suspensions (about 2 mg RNA/0.2 ml) were used as absorbance markers in co-sedimentations either with the VRs or with corresponding CRs. In the case of VRs, the 20,000 x *g* supernatants of DOC-treated virus were mixed with liver CR-suspension (0.2 ml) and run together on sucrose gradients. Specific activity of labelled myeloblastic CRs, isolated from cells labelled for 24 hr with $^3$H-uridine (see p. 101) was 6.2 x 10$^5$ cts/min/mg RNA. Sedimentation analysis of unlabelled myeloblastic CRs was carried out with a Spinco Model E with ultra-violet optics and an An-D rotor. CRs suspension, 0.800 A$_{260}$ units/ml of AMT, was centrifuged at 31,410 rev/min, 14 min and 20°C. Pictures were taken at 2 min intervals. S$_{20,w}$ values obtained in two separate analyses were 77.8 and 78.0.

**Sedimentation Characteristics of Ribosomes of 5-hr Old Virions and of Virus of Leukaemic Blood Plasma**

*Virus isolation*: 75 ml of fresh heparinized leukaemic chicken plasma (6 x 10$^{11}$ particles/ml) was mixed with 5 ml tissue culture supernatant containing 5-hr old virions produced by cells which had been labelled for 24 hr with $^3$H-uridine (see p. 101). Virus collection, purification and VR isolation, as well as absorbance and radioactivity determination, were as described on p. 100.

## Sedimentation Pattern of Virus Ribosome Nucleic Acids

*Virus isolation*: Virus from 260 ml of leukaemic chicken blood plasma was pelleted on a 50% sucrose gradient in an MSE rotor No 59100 at 13,000 rev/min, 90 min and +4°C. Collected virus suspension was diluted with water to 12% sucrose (5 ml end volume of sample) and sedimented on a discontinuous gradient consisting of 23% sucrose with 1/2 AMT (10 ml) and 17% sucrose with 0.005 molar EDTA, pH 7, (10 ml), in a Spinco rotor SW 25.1 at 25,000 rev/min, 90 min and +4°C. Isolation of VRs was accomplished as described on p. 101.

*RNA isolation*: VR-suspension in AMT supplemented with polyvinylsulphate (10 μg/ml) was lysed with 2% SDS in the presence of diethyl pyrocarbonate (10 μl/ml). Then the sample was extracted twice with phenol as described previously [10] and the extracted RNA was precipitated with ethanol [10]. Myeloblastic labelled rRNA was extracted from the isolated $^{32}$P-labelled myeloblast CRs (see p. 100) in the same way.

## Electrophoretic Pattern of Virus Ribosome Nucleic Acids

Electrophoresis was accomplished in polyacrylamide gels prepared according to Peacock and Dingman [15], containing 2.4% acrylamide and 0.5% agarose. Gels were run in the E buffer of Loening [16] supplemented with 0.2% SDS. The pre-running time was 15 min, and the run 3 hr at 4°C at a potential gradient of 8.0 V/cm and 5.0 mA/gel. RNA amounts applied per gel were 10 μg and 13.5 μg for VR-RNAs and myeloblast CR-RNAs, respectively. The mixture of both ribosomal RNA types consisted of 10 μg of VR-RNAs and 7.5 μg of myeloblastic CR-RNAs per gel. Samples stabilized in 2 molar acetic acid were washed in distilled water for 30 min and scanned at 260 nm in a gel UV-spectrophotometer constructed in this Institute by Dipl. Ing. J. Jirmus from the Electronics Group.

## RESULTS

### Time Conditions Optimal for Labelling of "Virus Ribosomes"

The pre-requisite for the detection of the "virus ribosomes" (VRs) was the determination of the conditions optimal for the labelling of the rRNA-like RNAs of the AMV. Such conditions arose from the pulse-labelling technique used, which made it possible to determine the time sequences of labelling of various nucleic acid species present in AMV relative to the virus rRNA-like RNAs.

Figure 1 shows the radioactivity sedimentation profiles of the virus nucleic acids and the absorbance profiles of the marker myeloblastic RNA obtained by glycerol density gradient centrifugation. The virus nucleic acids were extracted from the virions produced in successive 90 min periods by cells which were previously pulse-labelled for 60 min with $^3$H-Thd and then for another 60 min with $^{14}$C-Cd. As shown in this figure the $^{14}$C-radioactivity in the viral 65S RNA appeared very early, e.g. in the first 90 min after the corresponding pulse labelling, in accordance with the data available for the viral 70S RNA of the murine leukaemia virus [9]. On the contrary, the $^{14}$C-radioactivity of the "non-viral" RNAs, like the rRNA-like RNA species and the virus 4S RNA, started to appear only after a 450 min period from the end of the pulse-labelling with $^{14}$C-Cd.

This time-sequence of labelling, as determined for the "non-viral" RNAs, led to the assumption that the shortest time necessary for the appearance of the radioactivity in the presumed VRs should be 5-6 hr, as estimated from the end of the pulse labelling. At this time the radioactivity of the isolated presumed VRs should also be minimally contaminated with the radioactivity from the viral 65S RNA or its fragments. As shown in Fig. 1, the radioactivity of the 65S RNA constantly declined, starting from the 330 min period.

As for the comparison of the time-sequence of labelling of the virus rRNA-like RNAs, and of the appearance of the radioactivity in the virus-associated Thd-labelled material, most of the $^3$H-Thd-labelled material, identical with the 7.5S DNA of the AMV [7, 17] or RSV [18], began to be labelled earlier than the virus rRNA-like RNAs. The $^3$H-Thd-radioactivity in the 7.5S DNA began to appear during the 180 min period of Fig. 1, e.g. after 270 min from the end of pulse labelling with $^3$H-Thd, and increased constantly and appreciably during the whole period observed.

A minor part of the Thd-label was found to be associated, as in RSV [18], with the 65S RNA, revealing the same labelling pattern. When dissociated from the 65S RNA by RNAse treatment, this material sedimented at 5S. It was completely alkali-resistant, partially sensitive to DNAse (up to 30%), but only after a previous alkali treatment. It did not exhibit any changes in its sedimentation pattern when thernally denatured (not shown).

**Labelling Kinetics of the VR Structures**

In order to confirm the prediction concerning the time conditions optimal for the labelling of VR, we analysed by sucrose density gradient centrifugation the $^{32}$P- or $^3$H-Ud-labelled material present in the 20,000 x *g* supernatant of DOC-treated virions produced during 5 hr periods by cells previously pulse-labelled either with $^{32}$P or with $^3$H-Ud for 2 or 4 hr, respectively.

As shown in Fig. 2, the radioactivity sedimentation profiles of this material, derived from virions at all the time periods analysed, exhibited a distinct radioactivity peak situated approximately in the sedimentation region of the cytoplasmic ribosomal monomers [19] of Eucaryotes [14]. The radioactivity sedimentation profiles of this material derived from the first 5 hr period, labelled either with $^{32}$P or $^3$H-Ud, exhibited, apart from the main peak, a second smaller one located at a higher density region of the sucrose gradient. As will be shown

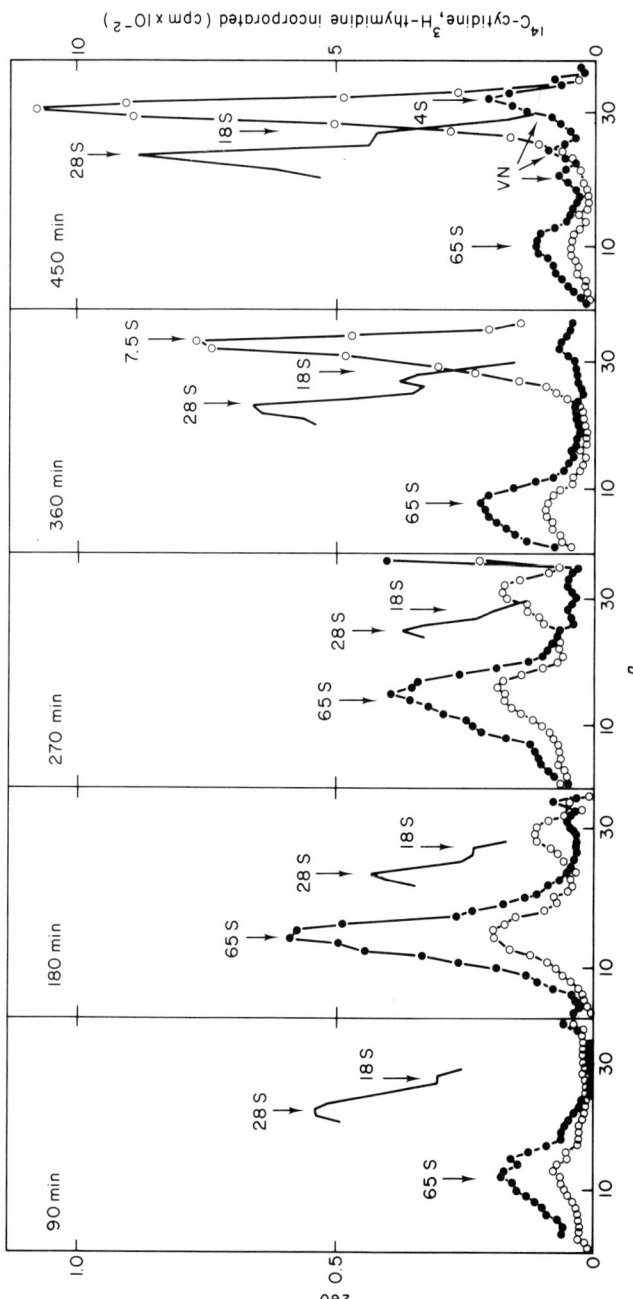

**Figure 1.** Radioactivity profiles of virus RNAs (●) and of virus Thd-labelled material (○) on a linear (40-10%) glycerol density gradient after 140 min at 37,000 rev/min and +4°C in a Spinco rotor SW 39.1, along with the absorbance profile of the unlabelled marker cytoplasmic RNA of leukaemic avian myeloblasts (———).

later, both observed radioactivity peaks corresponded to the VR-monomers (89S) and -dimers (135S).

The labelling characteristics of this virus material was in agreement with the prediction of conditions suitable for VR labelling. The radioactivity in the ribosomal sedimentation region appeared only after a 5 hr period from the end of the pulse-labelling with $^{32}$P or $^3$H-Ud. During an additional 5 hr it increased and subsequently decreased at a constant rate.

Analysing the 20,000 × $g$ supernatant of DOC-treated virions produced during consecutive 5 hr periods (5-15 hr) by cells constantly exposed to the

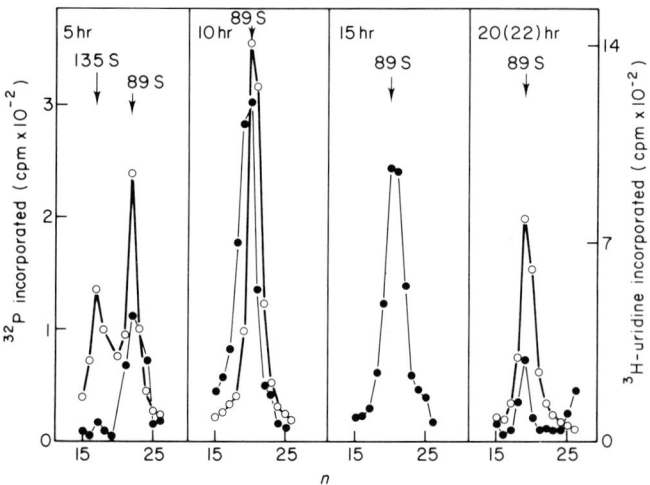

Figure 2. Radioactivity profiles of the VRs isolated from virus pulse-labelled either with $^3$H-uridine (o) or with $^{32}$P (•) and analysed on a linear (36-18%) sucrose gradient (supplemented with 1/2 AMT buffer), after 280 min at 25,000 rev/min and 24°C in a Spinco rotor SW 25.1.

radioisotope (not shown), a constant and linear increase of the radioactivity with the strong single peaks situated in the same sedimentation region (89S) was observed.

### Resolution of VR and Cytoplasmic Ribosomes by Sedimentation

In order to determine whether the VR and the cytoplasmic ribosomes (CRs) were identical, we examined by co-sedimentation in sucrose density gradients the behaviour of the labelled VRs and unlabelled CRs with a well-established value for the ribosomal monomer [20]. Figure 3A shows the sedimentation pattern of $^3$H-labelled VRs cosedimented with the unlabelled chicken spleen CRs. The radioactivity sedimentation profile of VRs exhibited a strong peak situated at about 89S and a smaller, but distinct, one at 135S when compared

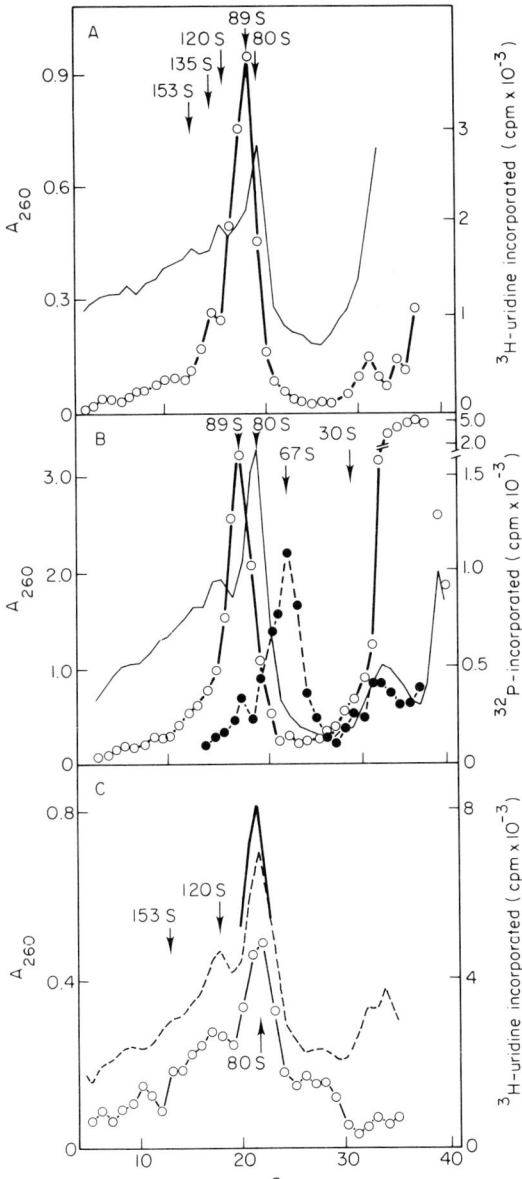

**Figure 3.** Co-sedimentation characteristics of VRs and CRs. (A) Radioactivity profile (o) of VRs labelled with $^3$H-uridine, and absorbance profile (———) of unlabelled chicken spleen CRs on a linear (36-18%) sucrose density gradient (supplemented with 1/2 AMT buffer), after 280 min at 25,000 rev/min and +4°C in a Spinco rotor SW 25.1. (B) Radioactivity profiles of VRs labelled with $^{32}$P (o) and of their dissociated monomers (●—●), and absorbance profile (———) of unlabelled chicken spleen CRs. (C) Radioactivity profile (o) of $^3$H-uridine-labelled myeloblastic CRs with absorbance profile of unlabelled chicken spleen CRs (– – –) and unlabelled chicken liver ribosomal monomers (———) on a linear (36-18%) sucrose density gradient after 130 min at 32,000 rev/min and +4°C in a Spinco rotor SW 39.1.

with the absorbance peaks of the spleen CR-monomer and -dimer which sedimented at S-values of 80 [20] and 120 [19], respectively. Figure 3B shows a similar sedimentation pattern for the $^{32}$P-labelled VRs cosedimented with unlabelled chicken spleen CRs. In this case the radioactivity profiles of VRs exhibited a strong peak at 89S and did not show the peak at 135S, but only a little shoulder in the region of 67S. When an aliquot of the 89S peak fraction was treated further, under conditions for ribosome dissociation [14] and once more analysed, there appeared, in addition to the substantially diminished radioactivity peak at 89S, a distinct peak in the region at 67S and two smaller shoulders, one at about 30S and a second at a lower S-value (Fig. 3B, broken line). By this behaviour the labelled VRs resembled in general the ribosomal structures, but they differed distinctly from the chicken spleen CRs by the S-values of their monomers and dimers as well as their sub-units.

Although an S-value of 80 should hold for all the eucaryotic CR-monomers [14], the absence of broader comparative data on S-values for CR-monomers of different chicken tissues and cells led us to compare the sedimentation patterns of the liver CRs with the CRs isolated from chicken spleen and from leukaemic myeloblasts which, in this case, were the virus producers. The comparison was accomplished by cosedimentation of $^3$H-labelled myeloblastic CRs with the unlabelled spleen or liver CRs in sucrose density gradients. Figure 3C shows that all the compared CRs exhibited the radioactivity or absorbance peaks of their ribosomal monomers at 80S. In addition to this comparison, the S value for the unlabelled myeloblastic CRs was determined with the analytical centrifuge, and gave an average value for the monomer of 78 (see p. 102). On the basis of these results, the VRs differed by their S-values also from the myeloblastic and spleen CRs. From the S-values reported for chicken mitochondrial ribosomes [20], they differ also from this type of ribosome.

### Basic Properties of VRs and of their RNAs

To further define VRs, it was necessary to extend the experimental data for the VRs isolated from the leukaemic chicken blood plasma virus (PV). The data obtained earlier in this paper were derived from the analyses of the VRs isolated from the 5-hr old virions produced by cells in tissue cultures (TC-V). Therefore it appeared reasonable, in the first instance, to compare the sedimentation characteristics of the VRs of the TC-V with those of the VRs isolated from the PV. In distinction to the TC-V, the PV represents a population of virions of much broader age differences. Figure 4 shows such a comparison achieved by co-sedimentation in sucrose density gradients of the labelled and unlabelled VRs of the TC-V. Both types were co-isolated from the mixture of the corresponding labelled and unlabelled virions. As shown in this Figure, the sedimentation profiles of the ribosome monomers of the TC-V and of the PV exhibited peaks situated at the same position, but they differed by the broader sedimentation profile of the ribosome monomers of the PV. This may imply a larger variation in the ribosome dimensions of the PV-VRs, most probably due to age differences when compared with the narrow sedimentation profile of the ribosome monomers of the 5-hr old virions. The value of the sedimentation coefficient $S_{20,w}$, obtained for the PV-ribosome monomers with the analytical centrifuge, was 85.4S (see p. 102). Micromorphologically, these VRs were ribosome-like. From their maximum linear dimensions of about 260 x 220 x 220 Å (unpublished results) they resemble the monomeric CRs of Eucaryotes [14].

As for the uv-optical properties, this material exhibited absorbance parameters comparable to the data given for the 80S CRs, possessing an RNA/protein ratio of about 1 [14]. The absorbance maximum and minimum were at

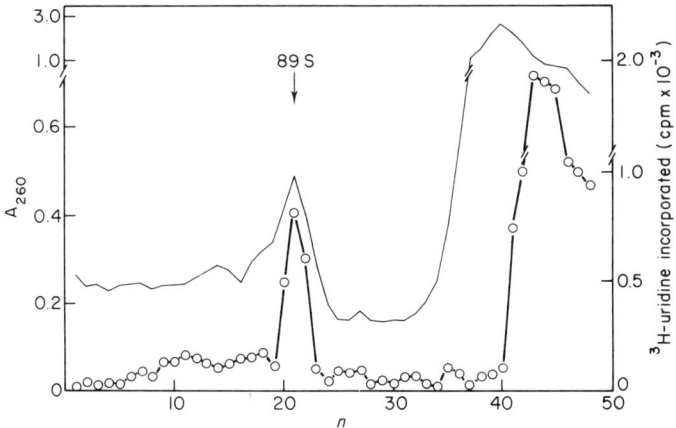

**Figure 4.** Radioactivity profiles (o) of $^3$H-uridine-labelled VRs of 5-hr old virions and absorbance profile (———) of unlabelled VRs of leukaemic chicken blood plasma virus co-sedimented on a linear (36-18%) sucrose density gradient (supplemented with 1/2 AMT), after 280 min at 25,000 rev/min and +4°C in a Spinco rotor SW 25.1.

260 and 235 nm, respectively; the $A_{260}/A_{280}$ and $A_{260}/A_{230}$ ratios were 1.51 and 1.48, respectively.

The nucleic acids extracted from the VRs were up to 98% RNase sensitive. When cosedimented in glycerol density gradients with $^{32}$P-labelled myeloblastic CR-RNAs, they exhibited, as shown in Fig. 5, two main components sedimenting at 29 and 19S. The absorbance profile of the VR-RNA also exhibited shoulders at 16 and 8S.

In order to test the validity of these small sedimentation differences observed between the VR- and the myeloblastic CR-RNAs, we analysed both types of RNA by electrophoresis on polyacrylamide gels using conditions [16] which permitted, on the one hand, a good resolution of the rRNAs and, on the other hand, a comparison of the separations achieved. Figure 6A represents the electrophoretic patterns of the myeloblastic CR-RNAs and of the VR-RNAs, run separately in duplicate at the same time, in adjacent tubes at the same voltage. The myeloblastic CR-RNA was separated into two main RNA species corresponding to the 28S and 18S rRNAs of the avian leukaemic myeloblasts. Beside these RNA-species a distinct amount of slowly migrating RNA, exhibiting a shoulder in the position of the 31S precursor RNA species [15], could be observed. By contrast, the V-RNAs resolved into four distinct components designed in Fig. 6A by the numbers 1-4. Component 1 consisted of multiple molecular RNA species mainly represented according to their mobilities by

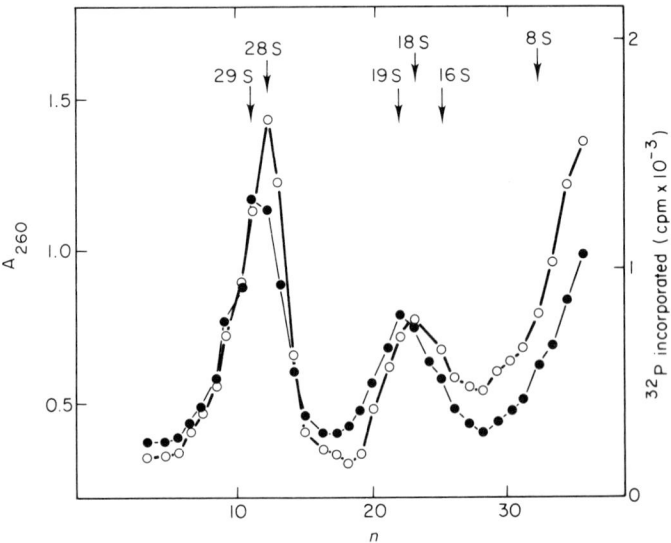

**Figure 5.** Radioactivity (○) and absorbance (●) profiles of $^{32}$P-labelled myeloblastic rRNA and unlabelled VR-RNA, co-sedimented on a linear (30-10%) glycerol gradient (supplemented with NaCl, 0.1 molar and Tris-HCl, 0.05 molar, pH 7.8) after 6 hr at 39,000 rev/min and +4°C in a Spinco SW 39.1 rotor.

molecules of molecular weight $2 \times 10^6$ (component 1b). Another RNA species present in the component 1 migrated like RNA with a molecular weight of about $2.4 \times 10^6$ [16] (component 1a).

The VR-RNAs present in component 1 resembled, by their mobilities, the molecular RNA precursors of the cytoplasmic 28S rRNA [21, 22]. The absorbance profile of component 1b also exhibited an inflexion at the position of the 28S RNA, suggesting also the presence of molecules of 28S RNA. The distinct VR-RNA components 2 and 3 represented RNA species migrating like molecules with molecular weights of $1 \times 10^6$ and $0.6 \times 10^6$, respectively, when compared with the mobilities of myeloblastic 18S rRNA with a presumed mol. wt of $0.7 \times 10^6$. The VR-component 4 migrated like RNA with a molecular weight of 0.3-0.4 x $10^6$. All these VR-RNA components represented the separated RNA species sedimenting in glycerol gradients at average S values of 19, 16 and 8, respectively. The component 2 resembled, according to its mobilities, the nuclear 21S precursor [21, 22] of the 18S rRNA. The VR-RNA components 3 and 4 may be identical with the postulated by-products of rRNA maturation, the polynucleotides $P_3$ and $P_2$, respectively [21]. As in the case of

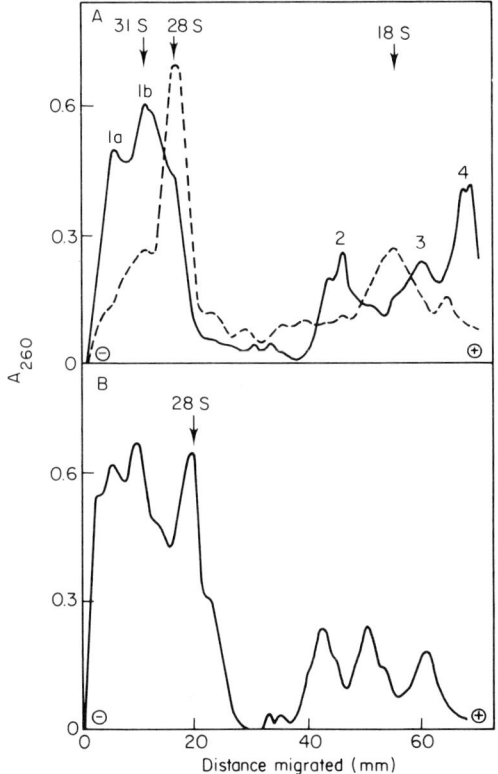

**Figure 6.** (a) Electropherogram of myeloblastic CR-RNAs (– – –) compared with that of VR-RNAs (———); (b) Co-electrophoresis of myeloblastic CR-RNAs with VR-RNAs.

the 28S RNA species which could be observed in smaller amounts in these VR-RNAs, an inflexion of the VR-RNA absorbance profile at the position of the 18S rRNA suggested the presence also of 18S RNA molecules.

To confirm these results, obtained by comparison of the two rRNA types separated individually, but under identical experimental conditions, we analysed the same rRNA types by coelectrophoresis [16] on the same polyacrylamide gels. Figure 6B shows the electropherogram of the myeloblastic CR-RNA and the VR-RNA run simultaneously in one tube. In accordance with expectations, the electropherogram of this RNA mixture exhibited the presence of the VR-RNA component 1, distinctly resolved from the myeloblastic 28S rRNA. There were present also three more rapidly migrating components located in the regions of the distances moved by the VR-RNA components 2-3 and by the myeloblastic 18S rRNA.

Summarizing the characteristics of the electrophoretic VR-RNA separations, we may conclude that the VR-RNAs were in this case mainly represented by the 31 and 21S precursor molecules of rRNAs. These findings are in agreement with the S-values obtained in this work for the VR-monomer and suggest a selective process of accumulation of these rRNA species in the virions. The fact that the VR-RNAs contain, on the one hand, RNAs resembling the postulated rRNA maturation intermediates [21], and on the other hand small amounts of the 28 and 18S RNA species, is suggestive of the processing of the rRNA precursor molecules which occurs directly in the virions during their ageing.

Such an interpretation is supported by our findings dealing with the sedimentation properties of the VRs isolated from the leukaemic chicken blood plasma virus, as described earlier in this paper. Accordingly a similar analysis of the VR-RNAs present in freshly produced virions would be expected to show the earliest stages and forms of the VRs and their RNAs.

## SUMMARY AND DISCUSSION

The detection of specific ribosomes in AMV raises several questions. With the 40 tRNA molecules present in AMV [2], these ribosomes—estimated as one per virion according to the "hypothetical virus BAI strain A composition" [2], and two per virion according to our evaluation, based on the different extractabilities of the individual virus RNA species (unpublished experiments)—may play a distinct role in protein synthesis in the virus-infected cells or in the virions. Expecially in the latter case, a residual and specific protein synthesis exerted by means of the virus-occluded cell enzymes might be necessary for the preservation or functioning of some virus components. The VRs might read the message either of a specific cell mRNA entering the virion by the same pathway as the VRs and tRNAs, or of a cistronic region of the proper viral RNA which is accessible to the VRs. Here, as well as in the virus infected cell, the VRs might have their special function, for example, in their different reading abilities in comparison to the CRs, as in case of the ribosome type specific reading of the mRNA observed in bacteria [23]. All these postulated events might take place in the region of the virus nucleoid where the presence of the ribosome-like particles has been demonstrated by electron microscopy in the case of the Bittner virus [24]. Finally, the detection of the VRs in AMV stresses again the importance of a knowledge of the molecular processes associated with the synthesis [25] and maturation [26] of the rRNAs in cells infected by the oncogenic RNA viruses.

The direct relation between cell rRNA processing and the virogeny, reflected even at the level of viral 65S RNA formation, has been shown only recently [10] in the case of AMV by means of toyocamycin, which is a specific inhibitor of rRNA maturation in mammals [27] and birds [26].

## REFERENCES

1. Beaudreau, G. S., Sverak, L., Zischka, R. and Beard, J. W., (1964). *Natn. Cancer Inst. Monogr.* **17**, 791.
2. Bonar, R. A., Sverak, L., Bolognesi, D. P., Langlois, A. J., Beard, D. and Beard, J. W., (1967). *Cancer Res.* **27**, 1138.
3. Obara, T., Bolognesi, D. P. and Bauer, H., (1971). *Int. J. Cancer* **7**, 535.
4. Carnegie, J. W., Deeney, A. O. C., Olson, K. L. and Beaudreau, G. S. (1969). *Biochim. biophys. Acta* **190**, 274.
5. Trávníček, M., (1969). *Biochim biophys. Acts* **182**, 427.
6. Trávníček, M. and Říman, J., (1970). *Biochim. biophys Acta* **199**, 283.
7. Říman, J. and Beaudreau, G. S., (1970). *Nature, Lond.* **228**, 427.
8. Erickson, E. L., (1969). *Virology* **37**, 124.
9. Bader, J. P. and Steck, T. L., (1969). *J. Virol.* **4**, 454.
10. Říman, J. (1971) *In* "The Biology of Oncogenic Viruses" p. 232, North-Holland, Amsterdam and London.
11. Adams, R. Z. P. and Lindsay, J. G., (1967). *J. biol. Chem.* **243**, 1314.
12. O'Brien, T. W. and Kalf, G. F., (1967). *J. biol. Chem.* **242**, 2172.
13. Küntzel, H. and Noll, H., (1967). *Nature, Lond.* **215**, 1340.
14. Spirin, A. S. and Gavrilova, L. P., (1969). "The Ribosome" pp. 20, 32, Springer-Verlag, Berlin, Heidelberg and New York.
15. Peacock, A. C., and Dingman, C. W., (1968). *Biochemistry* **7**, 668.
16. Loening, U. E., (1969). *Biochemistry* **113**, 131.
17. Fujinaga, K., Parsons, J. T., Beard, J. W., Beard, D. and Green, M., (1970). *Proc. natn. Acad. Sci. U.S.A.* **67**, 1432.
18. Levinson, W., Bishop, J. M., Quintrell, N. and Jackson, J., (1970). *Nature, Lond.* **227**, 1023.
19. Marbaix, G. and Burny, A., (1964). *Biochem. biophys. Res. Commun.* **16**, 522.
20. Rabbits, T. H. and Work, T. S., (1971). *FEBS Letters* **14**, 214.
21. Bush, H. and Smetana, K., (1970). "The Nucleolus" p. 257, Academic Press, New York and London.
22. Weinberg, R. A., Loening, U., Willems, M. and Peuman, S., (1967). *Proc. natn. Acad. Sci. U.S.A.* **58**, 1088.
23. Lodish, H. F., (1969). *Nature, Lond.* **224**, 867.
24. Gay, F., Clarke, J. K. and Dermott, E., (1970). *J. Virol.* **5**, 801.
25. Suskind, R. G., Pry, T. W. and Rabotti, G. F., (1969). *Cancer Res.* **29**, 1598.
26. Říman, J., Sverak, L., Langlois, A. J., Bonar, R. A. and Beard, J. W., (1969). *Cancer Res.* **29**, 1707.
27. Tavitian, A., Uretzky, S. G. and Acs, G., (1968). *Biochim biophys. Acta* **157**, 33.

## DISCUSSION

**M. S. van Dyk-Salkinova:** Dr. Říman, did you look for ribosomal protein in those particles from AMV which you did find to contain ribosomal RNA? Or do you think that these dense RNA-containing particles resemble rather the so-called informosomes which are claimed to contain only one type of protein? (Spirin *et al.*).

**J. Říman:** At the moment we have no information on this, but this aspect forms an integral part of our experimental programme designed for the elucidation of the nature and function of these particles.

# Reverse Transcriptase in Oncogenic RNA Viruses[*]

S. SPIEGELMAN and J. SCHLOM

*Institute of Cancer Research, Columbia University
College of Physicians and Surgeons, New York, New York, U.S.A.*

The DNA provirus hypothesis proposed in 1964 by Temin [1] stipulated that RNA replication of the RNA tumour viruses takes place through a DNA intermediate. This hypothesis served to explain the following unique biological and biochemical features of infections with RNA oncogenic viruses: (a) the heritably stable transformation of normal cells induced with these viruses; (b) the apparent vertical transmission of high leukemia frequency in reciprocal crosses between high and low frequency strains of mice [2]; and (c) the requirement for DNA synthesis [3] in the early stages of infection. The Temin hypothesis invoked the existence of an enzyme that can carry out a reversal of transcription by catalysing the synthesis of DNA on an RNA template. Evidence for such an enzyme was presented independently by Temin and Mizutani [4] and by Baltimore [5] who found a DNA-polymerizing activity in both an avian and a murine RNA tumor virus. The enzyme was detected by the incorporation of tritium-labeled thymidine triphosphate ($^3$H-TTP) into an acid-insoluble product that can be destroyed by DNase. Maximum activity required the presence of all four deoxyriboside triphosphates and magnesium. The fact that the activity was inhibited by RNase implied that the RNA of the virion is necessary for the reaction. These findings were so clearly pregnant with implications for the molecular details of viral oncogenesis that we undertook the task of providing a quick confirmation and extension of them [6].

## PROPERTIES OF THE RNA-INSTRUCTED DNA POLYMERASE REACTION (RIDP)

The purification of viruses is an extremely important procedure for the observation of the RIDP reaction and should not be taken lightly. The methodology for these procedures has been previously published [6-8]. It

---

[*] This study was conducted under contract 70-2049 within the Special Virus-Cancer Program of the National Cancer Institute, NIH, PHS, and grant CA-02332.

should be noted that some preparations require pretreatment with non-ionic detergents, whereas others do not. This is probably a function of the age and mode of preparations of the virus. In general, we have found that those preparations that do not require detergent pretreatment to exhibit activity are severely inhibited by the 0.2% of the detergent required to disrupt the virions. We have so far failed to find any ability to incorporate ribonucleoside triphosphates.

Table 1 shows that the polymerase has a mandatory requirement for all four deoxyriboside triphosphates as well as magnesium and is inhibited by the ribonucleases A and $T_1$.

**Table 1.** Requirements of AMV-polymerase: 13 $\mu$g protein; 40 min; 390 cpm/pmole*

| Condition | $^3$H-TTP Incorporated CPM |
|---|---|
| 4 dNTP; 2mM DTT; 6mM $Mg^{++}$; 60 mM $Na^+$ | 1460 |
| $K^+$ for $Na^+$ | 2345 |
| $-$dCTP | 165 |
| $-$dGTP | 199 |
| $-$dATP | 28 |
| $-Mg^{++}$ | 86 |
| Complete plus 10 $\mu$g/ml each of $T_1$ and Pancreatic RNase A | 171 |

* A standard incubation mixture of 1 ml contains in $\mu$moles: 50 of Tris HCl at pH 8.3; 6 of $MgCl_2$; 40 of KCl; 2 of dithiothreitol; 0.8 each of the non-labeled deoxynucleoside triphosphates; and .04 of labeled nucleosidetriphosphates. In this experiment, $^3$H-TTP (390 cpm/pmole) was used. The incorporations noted represent those observed in 0.1 ml aliquots, corresponding to 13 $\mu$g of viral protein, in 40 min.

Virus particles suspended in 0.01 M Tris (pH 8.3) at 320 $\mu$g of viral protein per ml were preincubated 10 min at 0°C in the presence of 0.2% Nonidet P-40 detergent with dithiothreitol (30 mM). The virus was then added to a standard incubation mixture at a level of 130 $\mu$g/ml and incubated at 37°C for 40 min. The reaction was terminated with the addition of 0.5 ml water and 0.3 ml of TCA mixture (equal volume mixture of 100% TCA solution, saturated sodium orthophosphate and sodium pyrophosphate). After 10 min, the precipitable radioactivity was collected on a nitrocellulose filter, dried, and the radioactivity determined in a liquid scintillation counter using BBOT scintillation fluid.

## PHYSICAL AND CHEMICAL NATURE OF THE PRODUCT

The product of an extensive synthesis was isolated and purified by phenol extraction and subjected to various chemical, physical and enzymological tests [6]. In agreement with earlier observations, the acid-precipitable product can be degraded by deoxyribonuclease but not by ribonuclease, pronase or NaOH. To provide further evidence of its DNA nature, the density of the product was examined by equilibrium density centrifugation in $Cs_2SO_4$. In these, as in subsequent centrifugations, internal markers were included to identify the density regions of RNA and DNA. It is clear from Fig. 1 that the product is found in the DNA region of the density gradient. The synthetic DNA has a density of 1.450 g/cm³, somewhat heavier than the DNA prepared from mouse embryo fibroblasts (MEF-DNA) used as a density (1.420 g/cm³) marker.

The synthesis of even a proper DNA heteropolymer, however, does not establish its relevance to the life history of the virus. The central issue is whether

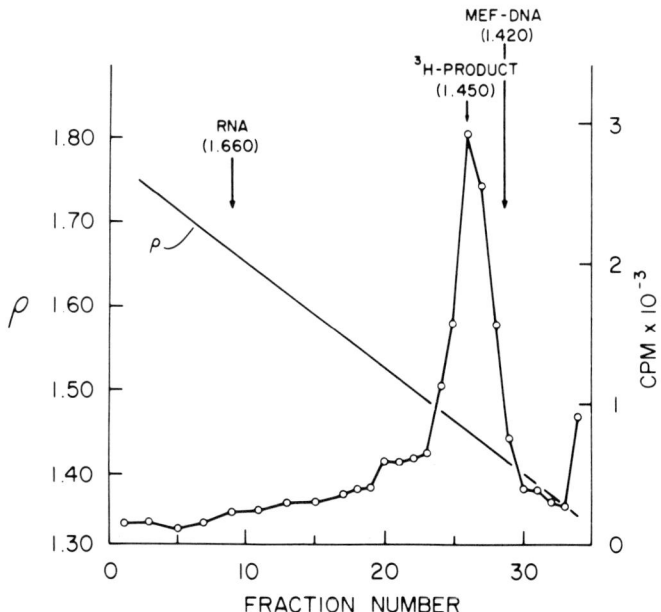

**Figure 1.** Density of product synthesized by RSV (RAV-1) polymerase. One ml of a standard reaction containing 130 μg of viral protein and 0.1 μM of ³H-TTP at 400 cpm/pmole were incubated for 8 hr at 37°C. The product was purified as described [6], mixed with saturated $Cs_2SO_4$ to a density of 1.550 g/cm, and centrifuged at 33,000 rpm at 20°C for 60 hr in a Spinco SW56 rotor.

the sequence of the synthetic DNA is related to the RNA of the virion. The most obvious relationship to expect is complementarity and we therefore resorted to molecular hybridization [9].

Because the system being studied was so novel, and to maximize the amount and certainty of the information obtained, equilibrium centrifugation in cesium salts was used to detect hybrid formation. $Cs_2SO_4$ was used to permit banding of both RNA and DNA as well as all intermediate hybrid structures.

Figure 2 shows the outcome of an annealing reaction between ³H-DNA synthesized with the Rauscher leukemia virus (RLV) polymerase and RNA purified from RLV virions. Here the viral RNA is present in great excess. It will be noted that virtually all the DNA has moved from the DNA region to the density position characteristic of RNA. This indicates that all the DNA synthesized in this reaction is complementary to the viral RNA.

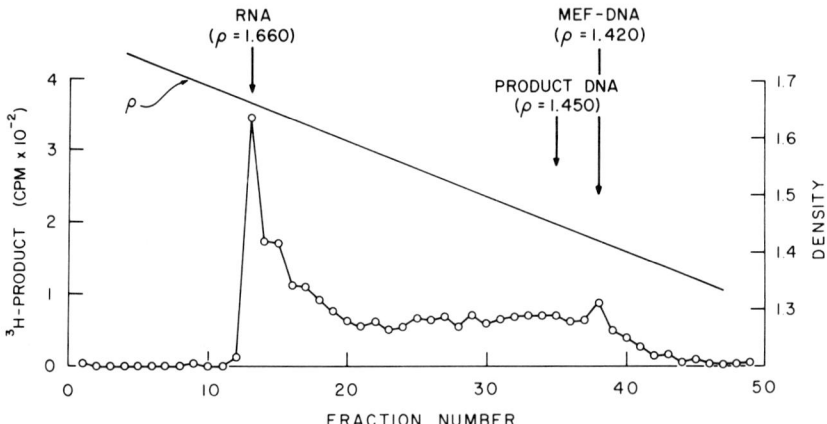

**Figure 2.** Cesium sulfate ($Cs_2SO_4$) equilibrium density gradient centrifugation of RLV-polymerase product after annealing to RLV viral RNA. A 5 ml standard reaction containing 340 μg of viral protein and $^3$H-TTP at $1.8 \times 10^3$ cpm/pmole was incubated for 2 hr at 37°C. The synthesized product was purified as described [6]. In addition, the product was subjected to digestion in 0.4 M NaOH for 24 hr at 37°C to remove RNA from pre-existing complexes, a step essential for interpretable annealing experiments.

Viral RNA was extracted from particles harvested from supernatants of the JLS-V5 mouse cell line growing in roller bottles. The particles were purified by differential centrifugation and banding in sucrose, RNA was extracted and purified with phenol banded in a glycerol gradient, and the 70S peak collected for use [6].

The purified product (5000 cpm) and 10 μg of 70S RLV-RNA were dissolved in 90 μl of EDTA (0.002 M at pH 7.4) and formamide (50% final concentration). The mixture was heated to 68°C for 10 min to denature any pre-existing structures in the RNA preparation. After quick chilling, 10 μl of 0.4 M NaCl was added. The mixture was then held at 37°C for 23 hr. This annealing method minimizes fragmentation of the RNA. After the annealing, the mixture was subjected to $Cs_2SO_4$ gradient centrifugation as described in the legend to Fig. 1.

## SURVEY OF RNA VIRUSES FOR RNA-INSTRUCTED DNA POLYMERASE

Table 2 summarizes the relevant available information concerning the presence of RNA-instructed DNA polymerase in oncogenic and non-oncogenic viruses. It will be noted that the 11 groups listed include viruses that cause leukemias, mammary carcinomas, or sarcomas in chicken, mice, rats, and monkeys. As may

**Table 2.** DNA polymerase activities of RNA viruses

| Virus | Source | Activity | References |
|---|---|---|---|
| *Viruses of Known Oncogenic Potential* | | | |
| Rous sarcoma | | | |
| (RAV-1) | CEF* | + | 6 |
| (Prague) | Cell culture (chicken) | + | 5 |
| (Schmidt-Ruppin, B-77) | Cell culture (chicken) | + | 4 |
| Avian leukosis (Mc 29) | CEF | + | 36 |
| Avian reticuloendotheliosis | | | |
| (Twiehaus agent) | CEF | + | 7 |
| Avian myeloblastosis | Chicken plasma | + | 6 |
| | Cultured myeloblasts | + | 7 |
| Murine leukemia | | | |
| Rauscher | Plasma | + | 5, 6 |
| Rauscher | MEF | + | 7 |
| Rauscher | JLS-V5 cell line | + | 7 |
| AKR | Cell culture (rat) | + | 37 |
| Moloney | JLS-V9 cell line | + | 38 |
| Murine sarcoma-leukemia complex | | | |
| Moloney | Mouse tumors | + | 6 |
| Moloney | MEF | + | 7 |
| Moloney | 78A1 rat cells | + | 7 |
| Harvey | MEH mouse cells | + | 39 |
| Kirsten | NRK cells | + | 38 |
| Murine mammary tumor | Paris R III milk | + | 6 |
| | $C_3H$ milk | + | 38 |
| Feline leukemia | | | |
| Ricard | Cell culture (feline) | + | 6 |
| Thielen | Cell culture (feline) | + | 7 |
| Gardner | Cell culture (canine) | + | 37 |
| Feline sarcoma | | | |
| Ricard | Cell culture (marmoset) | + | 7 |
| Gardner | Cell culture (feline) | + | 38 |
| Gardner | Cell culture (canine) | + | 37 |
| Hamster sarcoma | B34 hamster cells | + | 7 |
| Hamster leukemia virus | LSH hamster cells | + | 37 |
| Visna, progressive pneumonia | Sheep cells | + | 16, 7, 17 |
| *C-type and/or B-type Viruses of Unproven Oncogenicity* | | | |
| Mammary tumor | R35 rat cells | + | 25 |
| Mammary tumor (Mason-Pfizer) | Cell culture (monkey, human) | + | 6 |
| Viper | VSW cells | + | 37 |
| Human milk agent | Human milk | + | 8 |
| ESP-1 | Human cells | + | 40 |

Table 2—contd.

| Virus | Source | Activity | References |
|---|---|---|---|
| *Non-oncogenic Viruses* | | | |
| Newcastle disease | Allantoic fluid | — | 20 |
| Influenza (A and WSN) | Allantoic fluid | — | 20 |
| Reo | L-cells | — | 20 |
| Vesicular stomatitis | BHK cells | — | 5, 20 |
| Polio | HeLa cells | — | 20 |
| Sendai | Allantoic fluid | — | 38 |
| Respiratory syncytial | HEP-2 cells | — | 38 |
| Foamy | Monkey cells | + | 41 |

\* Abbreviations: CEF = chick embryo fibroblasts
MEF = mouse embryo fibroblasts

be seen from Table 2, the source or strain of the virus has no influence on the detectability of the enzyme. Viruses derived from tumors, plasmas of infected animals, or tissue cultures employing cells homologous or heterologous to the host of origin all possess RNA-instructed DNA polymerase activity. Note that there are two "exceptions" to the rule in that visna and foamy viruses possess the reverse transcriptase. Foamy—ubiquitous in monkeys—is thus far a "virus without a disease." Its oncogenic potential has not yet been thoroughly investigated and it is therefore difficult to draw conclusions at this point.

## THE "EXCEPTION" OF VISNA VIRUS

Visna virus is a neurotropic agent isolated from the brains of neurologically afflicted sheep [10, 11]. Clinical manifestations of visna infection are extremely similar to those of multiple sclerosis in man. Such neurological signs do not appear until many months or years after virus inoculation, following a disease pattern typical of "slow" virus infection [12, 13]. Once clinical signs appear, the disease progresses relentlessly to paralysis and death.

Visna virus can be propagated *in vitro* in primary tissue cultures derived from a variety of sheep tissues, including choroid plexus and testis. Studies of virus replication in cell culture have shown that visna virus resembles the tumor-forming RNA viruses in morphology, morphogenesis and certain physical properties. Furthermore, like the oncogenic RNA viruses, it is sensitive to such agents as actinomycin D and bromodeoxyuridine, which interfere with DNA synthesis [14,15].

Visna virus contains the RNA-instructed DNA polymerase activity [7, 16, 17] as well as DNA polymerase activities that respond to DNA as well as DNA-RNA hybrid duplexes [7].

We have shown [18] the major nucleic acid component recovered from visna virions to have a sedimentation coefficient of approximately 67S. In addition, a slower sedimenting moiety of approximately 5 to 7S was regularly detected. The major 60-70S component was shown to be single-stranded RNA by its complete sensitivity to RNase and density after isopycnic gradient centrifugation of cesium sulfate. The slowly sedimenting virus 5-7S nucleic acid species was not completely digested by treatment with RNase. On isopycnic centrifugation with cesium sulfate, the 5-7S species contained components that band at the density of RNA as well as components that band at density characteristic of RNA . DNA hybrid complexes. Thus, the RNA species extracted from visna virions resemble those which have been recovered from the oncogenic RNA viruses.

These studies indicate that visna virus contains nucleic acids and DNA polymerase activities similar to those found in tumor-forming RNA viruses. This, coupled with a recent report [19] that visna virus is able to cause transformation of murine cell cultures, which then cause tumors in mice, indicates that visna is one of a group of slow agents that share common properties with the RNA tumor viruses and have oncogenic potential. By utilizing the available information about the biological and chemical properties of visna virus, it may be possible to incriminate similar cryptic agents as causes of currently undefined human neurological diseases.

## DNA-DIRECTED DNA POLYMERASE ACTIVITY IN ONCOGENIC RNA VIRUSES

The heritably stable state that characterizes cells transformed by these oncogenic agents seems to require integration of the newly synthesized DNA into the genome of the cell. It is unlikely that the RNA-DNA hybrids detected as presumptive intermediates in the polymerase reaction can serve this purpose. It is more likely that the single-stranded DNA has to be converted into its double-stranded equivalent for integration, a conversion that would necessitate a DNA-directed DNA polymerase. We therefore pointed to the need to search for such an enzyme, either in the virion or in the infected cell, which would presumably be characterized by its ability to use duplexes to generate duplexes. We have examined numerous oncogenic RNA viruses, including Rauscher leukemia virus (RLV), Rous sarcoma virus RSV-(RAV-1), avian myeloblastosis virus (AMV), murine mammary tumor virus (MTV), Moloney sarcoma virus (MSV), and the feline leukemia virus (FeLV), and encountered a DNA-directed DNA polymerase in the virions of all of them [20].

The addition of either *Escherichia coli* DNA or mouse embryo fibroblast (MEF) DNA into a polymerase reaction mixture results in a striking stimulation of $^3$H-dATP incorporation into DNA. A trivial explanation of this would invoke protection by the added DNA against nucleolytic degradation of the DNA synthesized by the RNA-directed polymerase. To test—and eliminate—this, and at the same time provide a more convenient system for studying the DNA-directed step, it was necessary to eliminate the RNA-directed activity. This can be done by destroying the resident viral RNA by previous treatment of the disrupted virions with a suitable nuclease. Micrococcal nuclease is convenient because it requires $Ca^{2+}$. Its activity can consequently be readily neutralized by the specific chelating agent, ethyleneglycol-*bis*-(aminoethyl ether) tetra-acetic acid (EGTA). There is little residual activity in the absence of added exogenous nucleic acid. The DNA-stimulated reaction requires all four deoxyriboside triphosphates as well as $Mg^{2+}$.

The product of an extensive synthesis was purified by phenol extraction and alcohol precipitation. It was then subjected to physical and enzymological tests. The acid-precipitable product can be degraded by deoxyribonuclease and spleen phosphodiesterase but not by ribonuclease or alkali. To confirm that it is DNA, the density of the product was examined by equilibrium density centrifugation in $Cs_2SO_4$. The density of the DNA synthesized in response to the addition of MEF DNA is found at the same density (1.420) as MEF DNA. The DNA produced by the RNA-directed reaction has the indicated density of 1.450.

The average size of the product was compared with that of template in neutral and alkaline sucrose gradients. It is clear that the radioactive product is smaller than the DNA used to stimulate the reaction; there is no evidence of covalent attachment of the product to the template.

Table 3 compares the response of the AMV DNA-DNA polymerase to different DNA templates. Unlike the thoroughly studied DNA-dependent DNA polymerase of Kornberg [21], these oncogenic viral DNA polymerases prefer double- to single-stranded DNA. A particularly interesting example is the fl DNA, a single-strand DNA quite similar to that found in the DNA bacteriophage φX174, which has been shown [22] to be an excellent template for the Kornberg enzyme. We have seen no signs of template function even in extensive incubations with fl DNA. The fl-RF DNA, a covalently linked double-stranded circular DNA, is also inactive. This is not surprising because at least one of the two strands would have to be nicked before replication could commence.

The response to quite disparate DNA templates provides an excellent opportunity of deciding by molecular hybridization whether the DNA added is in fact serving as an instructive agent in the polymerization. The required DNA-DNA hybridization can readily be performed using the Denhardt [23] modification of the Gillespie and Spiegelman [24] method for hybridizing RNA to DNA fixed to membrane filters. Table 4 shows the results of hybridizing each

**Table 3.** Response of AMV DNA-DNA polymerase to a variety of DNA templates (cpm $^3$H-TTP incorporated).*

| DNA | None | Double-stranded | Single-stranded or denatured |
|---|---|---|---|
| E. coli | – | 1006 | 239 |
| MEF | 135 | 638 | 135 |
| CEF | 76 | 889 | 403 |
| T6 | 120 | 2320 | 460 |
| f1 | 50 | – | 19 |
| f1-RF | 50 | 128 | – |

* The reactions were carried out under the standard conditions and processed as described [20]. The $^3$H-dATP had a specific activity of 350 cpm/pmole. Incubations were for 30 min at 37°C. DNA concentrations were about 3 µg/0.25 ml.

**Table 4.** Hybridization of DNA-directed products to various DNAs*

| DNA on Filter | Template for Synthetic DNA | | |
|---|---|---|---|
| | Mouse | E. coli | T6 |
| Mouse | 8698 (100%) | 587 (6%) | 1,477 (2%) |
| E. coli | 454 (5%) | 7724 (100%) | 197 (0.4%) |
| T6 | 152 (2%) | 168 (2%) | 54,673 (100%) |

* Radioactive material in each hybridization was equivalent to $3.3 \times 10^4$, $4.9 \times 10^4$, and $9.4 \times 10^4$ cpm for DNAs synthesized with mouse, E. coli, and T6 DNA, respectively. The numbers represent cpm found per 150 µg of DNA on the filter; those in parentheses represent percentage of that observed in the homologous hybridization.

Products were prepared as described [20], and dissolved in 0.5 ml 0.07 x SSC (1 x SSC contains 0.15 M NaCl and 0.015 N sodium citrate, pH 7.0).

The hybridization reactions were carried out using DNA immobilized on membrane filters and product DNA in solution in 200 µl of 3 x SSC at 66°C for 12 hr. For denaturation, 100 µl of the product was made 0.1 M with respect to KOH and kept at room temperature for 10 min. The solution was then neutralized with an equivalent amount of HCl and diluted to a final volume of 200 µl with a calculated amount of SSC solution so that the final solution was 3 x SSC. The filters were loaded with 100 to 150 µg of denatured DNA, allowed to dry at room temperature, and then transferred to a vacuum oven at 80°C for 2 hr at a pressure of 25 mm Hg. The filters were pre-incubated in Denhardt [23] solution (0.02% bovine serum albumin, 0.02% "Ficoll" and 0.02% polypyrrolidene in 3 x SSC) at 66°C for 6 hr immediately preceding hybridization. After hybridization for 12 hr, the filters were taken out; each was washed on either side with 100 ml of $10^{-3}$ Tris, pH 9.3, dried and counted.

of three DNA products to the three DNA templates used in the synthetic reactions. The data are clear cut: the hybridizability of each product is much superior when challenged with the DNA actually used in its synthesis. The fact that the DNA-directed polymerases described here readily accept, indeed prefer, double-stranded DNA as templates distinguishes them from the DNA-dependent DNA polymerases so far reported.

All the oncogenic RNA viruses that we have examined possess the DNA-directed polymerase, but this activity could not be detected in five non-oncogenic viruses: reo, polio, influenza, vesicular stomatitis, and Newcastle disease virus [20].

## SYNTHETIC DNA-RNA HYBRIDS AND RNA-RNA DUPLEXES AS TEMPLATES FOR THE POLYMERASES OF THE ONCOGENIC RNA VIRUSES

We were still left with the problem of identifying the polymerization reaction that converts the DNA-RNA hybrids into the DNA-DNA duplex. To synthesize natural DNA-RNA hybrids in amounts adequate for enzymatic test is still a formidable task. To avoid the logistic problems involved, we decided to use synthetic polynucleotide duplexes. We have found that certain of the synthetic duplexes are not only functional, but are superior to natural templates by almost two orders of magnitude in stimulating polymerization [25]. This finding leads to an extraordinarily useful tool for detecting the enzyme activity. It is important to note, however, that the response of an enzyme to synthetic duplexes does not define it as a reverse transcriptase. For example, the DNA polymerase of *E. coli* has been shown to use the synthetic ribopolymer rA:rU to synthesize deoxyribopolymer dA:dT [26].

Table 5 illustrates the responses of the six RNA oncogenic viral preparations to the resident RNA templates (column 1), exogenous DNA (column 2), and to the hybrid duplex (column 3) formed from polydeoxycytidylate and polyriboguanylate (dC . rG).

For all six viruses there is clear evidence of a hybrid-directed DNA synthesis that is dramatically superior to the reactions previously studied.

The abilities of various synthetic double-stranded DNA and RNA homopolymers to serve as templates are recorded in Table 6. Except for rC:rG, all stimulate some incorporation of one of the complementary nucleotide pairs. Of those tested, rI:rC is clearly the best, followed by dC . dG and dC . dI. It is interesting to note that, with the exception of rC . rG, the polyribopolymers are superior to the corresponding deoxypolymers as templates. The template capabilities of a variety of synthetic DNA-RNA hybrid duplexes were also examined. Some of these are the best templates, notably dC . rG and dI . rC.

A detailed examination of the intermediates in the reactions mediated by these synthetic duplexes should be very useful for the study of the chemistry of the polymerization and the factors that determine its asymmetry. Even a casual inspection of Table 5 reveals the obvious advantage of using templates such as dC.rG for detecting enzyme in cells or virus particles and following activity

Table 5. Responses of oncogenic viral DNA polymerase to added DNA and a synthetic RNA-DNA hybrid. Numbers are pmoles of nucleotide incorporated in 10 min per 10 $\mu$g of viral protein*

|  | Template Added | | |
| --- | --- | --- | --- |
| Polymerase | None | CEF-DNA | dC:rG |
| AMV | 2.8 | 5.3 | 140.0 |
| FeLV | 2.1 | 2.6 | 30.7 |
| MSV | 1.2 | 1.2 | 30.0 |
| RLV | 1.5 | 4.3 | 162.0 |
| MTV | 0.1 | 1.0 | 35.1 |
| R-MTV | 0.1 | 0.6 | 21.4 |

* A standard incubation mixture of 0.25 ml contains in $\mu$moles: 12.5 of Tris-HCl (pH 8.3); 3 of $MgCl_2$; 10 of KCl; 2.5 of dithiothreitol; 0.04 of each of the deoxyribonucleoside triphosphates. To monitor the reaction, $\alpha$-$^{32}$P-dGTP was used at a specific activity of 65 cpm/pmole. Reactions that were stimulated by added templates were carried out by including the template at a level of 1.0 $\mu$g per 0.25 ml in the standard incubation mixture. The incorporations noted above represent those observed in 50 $\mu$l aliquots corresponding to 10 $\mu$g of viral protein.

Virus particles suspended in TNE buffer (Tris 0.01 M, pH 8.3; NaCl 0.1 M; EDTA 0.002 M) at a concentration of 2.3 mg of viral protein per ml were preincubated for 10 min at 0°C in the presence of 0.2% "Nonidet P-40" and 0.1 M dithiothreitol. The incubated solution was then added to a standard incubation mixture at a level of 50 $\mu$g of viral protein per 0.25 ml and incubated at 37°C. After 10 min of incubation, 50 $\mu$l aliquots were withdrawn and processed for determining the acid-precipitable radioactivity [25]. The particles designated here as R-MTV were obtained from supernatant fluids of rat mammary tumor, R-35 grown *in vitro* at Charles Pfizer Co., Maywood, New Jersey. The source of the other viruses and their methods of purification have been described previously [25]. R-MTV was purified in the same way as the monkey mammary tumor virus.

during purification procedures. We have now established that oncogenic viruses contain DNA polymerase activities directed by single-stranded RNA, double-stranded RNA, double-stranded DNA, and DNA-RNA hybrids. Further, all oncogenic RNA viruses examined contain all these activities.

Table 6. Response of AMV-polymerase to synthetic double-stranded DNA and RNA polymers*

| Template | Incorporation (pmoles) 5 min/10 μg viral protein | |
|---|---|---|
| rA : rU | dT (15.0) | dA (0.4) |
| rC : rG | dG (< 0.1) | dC (< 0.1) |
| rI : rC | dG (128.0) | dC (2.1) |
| dA : dT | dT (0.8) | dA (1.4) |
| dC : dG | dG (26.9) | dC (0.4) |
| dC : dI | dG (24.9) | dC (2.0) |

* The reaction mixtures and conditions of reactions were the same as described in the legend to Table 5. The duplexes were prepared by using the appropriate polymer components, using the method described for dC:rG (see legend to Table 5).

## DISTINGUISHING REVERSE TRANSCRIPTASE OF AN RNA TUMOR VIRUS FROM OTHER KNOWN DNA POLYMERASES

We have recently [27] purified the reverse transcriptase from the avian myeloblastosis virus and have shown that it possesses the RNA, DNA, and synthetic hybrid-directed DNA polymerase activities observed in crude preparations of disrupted virions. The availability of purified "reverse transcriptase" makes it possible to compare its properties with the well known cellular DNA polymerases described by Lehman *et al.* [28] and Bollum [29], and thereby to identify any features that distinguish the reverse transcriptase [30].

Table 7 compares the activities exhibited by the three polymerases with six different templates. It is immediately apparent that all three enzymes respond to the three synthetic duplexes and to the natural double-stranded DNA. However, only the AMV reverse transcriptase exhibits activity with AMV and Qβ single-stranded RNA [30].

The data summarized in Table 7 also show that DNA and the synthetic duplexes can all be used as templates by all three enzymes. It is obvious, therefore, that none of them *alone* can serve to distinguish reverse transcriptase from the two cellular polymerases. It is equally obvious that the two viral RNAs readily identify the reverse transcriptase as the only one of the three DNA polymerases that can use RNA as template.

The results with the oligomer-homopolymer complexes also provide a distinction between a reverse transcriptase and a normal cellular polymerase. A reverse transcriptase greatly prefers initiator-primed synthesis with $(dT)_{10} \cdot poly(A)$ as a template, whereas the cellular polymerases give reactions with $(dT)_{10} \cdot poly(dA)$ that are either greater or equal to those with $(dT)_{10} \cdot poly(A)$, depending on the purity of enzyme preparation. Adherence to the criteria and precautions outlined should alleviate some of the confusion that has been generated by an uncritical dependence on the sole use of synthetic duplexes, such as poly(A) . poly(dT) and poly(A) . poly(U), for the detection of reverse transcriptase.

Table 7. Comparison of template responses*

| Template | Polymerases | | |
|---|---|---|---|
| | AMV | E. coli | Calf thymus |
| poly(dA).poly(dT) | 28.0 | 96.0 | 107.0 |
| poly(A).poly(dT) | 81.0 | 21.0 | 30.0 |
| poly(A).poly(U) | 27.0 | 4.4 | 15.0 |
| ML-DNA | 1.4 | 2.9 | 0.9 |
| AMV RNA | 1.5 | <0.02 | <0.02 |
| Qβ RNA | 0.4 | <0.02 | <0.02 |

\* Amount of $^3$H-TTP incorporated into acid-insoluble product in a 20-minute assay with the AMV polymerase, E. coli polymerase, and calf thymus polymerase. The reaction mixtures and conditions were as described [30]. For reactions involving synthetic duplexes, the conditions were altered as follows: (a) only one triphosphate, TTP, was present in assays involving synthetic duplexes, (b) the concentration of labeled TTP was ten times less so that the specific activities were approximately 50 cpm/pmole, and (c) the synthetic duplexes were used at a concentration of 10 μg/ml. The synthetic duplexes poly(dA) . poly(dT) and poly(A) . poly(U) were prepared by annealing equivalent amounts of the appropriate homopolymers at room temperature in 0.1 M NaCl.

## PURIFICATION OF THE DNA POLYMERASE OF AVIAN MYELOBLASTOSIS VIRUS

Polymerase from avian myeloblastosis virus has been purified by a combination of column chromatography and gel filtration methods [27]. The isolated enzyme sediments at approximately 6S and consists of two subunits of molecular weights 110,000 and 69,000. It is free of RNA and DNA endonuclease activity. The enzyme possesses the RNA-, DNA-, and hybrid-directed polymerase activities found in the virion.

## AGENTS ASSOCIATED WITH MAMMARY CARCINOMAS OF MICE, MONKEYS AND HUMANS

Our observation [6] of the reverse transcriptase in the mouse mammary tumor virus (MMTV), or Bittner virus, was one of the most interesting to date, in our opinion. This agent differs in morphology (type-B), protein content, antigenicity, and density from the known "type-C" RNA oncogenic virus that causes leukemias and sarcomas in both fowl and mammals. The presence of the reverse transcriptase and a high molecular weight (60S-70S) RNA appear to be diagnostic for oncogenic potential in the RNA viruses.

## MASON-PFIZER MONKEY VIRUS (M-PMV)

The presence of a virus, morphologically similar to that of the known RNA tumor viruses, has been reported recently in a spontaneous mammary tumor of a rhesus monkey [31]. These particles were initially propagated by co-cultivation of the mammary tumor tissue with monkey embryonic cells and they have since been grown in a variety of monkey, chimpanzee, and human cell lines [32]. This virus has recently been designated the Mason-Pfizer monkey virus (M-PMV) [33].

Virions isolated from a spontaneous mammary carcinoma of a rhesus monkey and propagated in human cells possess an RNA-instructed DNA polymerase. They also exhibit DNA polymerase activities that respond to either double-stranded DNA or synthetic RNA–DNA hybrid complexes as templates [34]. The virion has a density of 1.16 g/ml and contains a nucleic acid species of high molecular weight (sedimentation coefficient, 60-70S), which bands as RNA at 1.670 in a $Cs_2SO_4$ equilibrium density gradient. In addition, the virions contain species of low molecular weight (4-6S) that consist of RNA as well as components banding at densities characteristic of DNA–RNA complexes. The nucleoid of this virion has been isolated and shown to have a density of 1.23 g/ml; it also contains a 60-70S nucleic acid species [34].

## RNA-INSTRUCTED DNA POLYMERASE ACTIVITY IN VIRUS-LIKE PARTICLES ISOLATED FROM HUMAN MILK

Particles morphologically identical to the type B mouse mammary tumor virus have been observed in human milk and in human mammary carcinomas. In a recent epidemiological survey by Moore et al. [35] such particles were readily detected in the milk of 60% of American women with a familial history of breast cancer. In American women with no such familial history, particles were found with a frequency of only 5% and when detected they were usually present in much lower amounts. They were also found in considerable quantities in 39% of

Parsi women of Bombay, a group that has a two to three times higher incidence of breast cancer than the rest of the Bombay population. Breast cancer accounts for about half of all cancer in Parsi women. The Parsis are descendants of Zoroastrians who emigrated from Persia some 1300 years ago and who, because of their strict religious rules, have been inbreeding ever since.

Some of the milks contained particles in quantities sufficient to permit isolation and examination of their enzyme content. We reported [8] a survey of

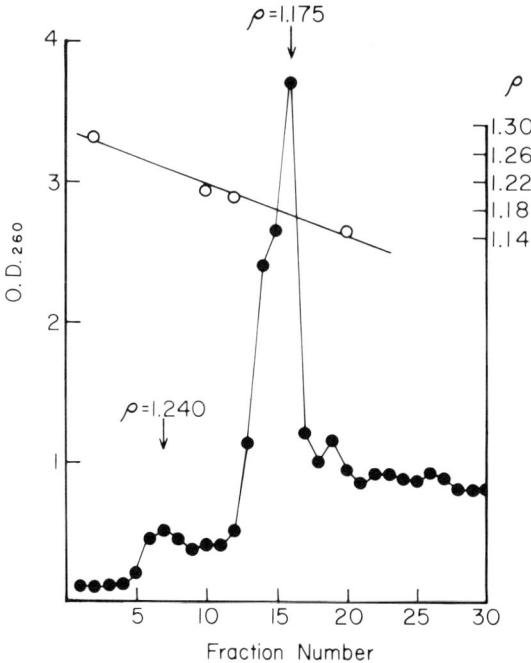

**Figure 3.** Sucrose equilibrium gradient centrifugation of human milk particles. 125 ml of milk positive for type-B particles were purified as described [8].

thirteen human milks: all milks positive for type B particles by electron microscopy contain particles that band at densities identical to those of the known oncogenic RNA viruses (1.16-1.19 g/ml [Fig. 3]) and exhibit the RNA-instructed DNA polymerase, or "reverse transcriptase," characteristic of these tumor viruses (Fig. 4). All milks negative for particles were also devoid of detectable reverse transcriptase.

Although the possession of an RNA-instructed DNA polymerase in purified particles is by no means proof of oncogenic potential, this enzyme activity has been observed in all isolates of oncogenic RNA viruses (Table 2).

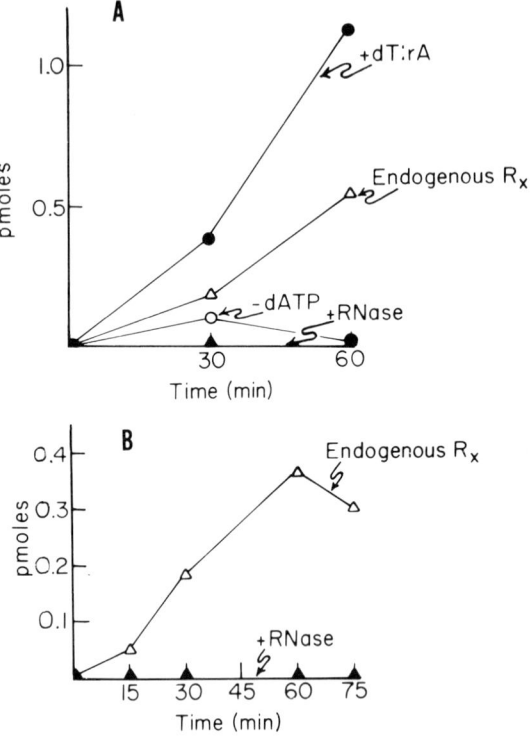

**Figure 4.** Characterization of RNA-instructed DNA polymerase activity of human milk particles. (a) Illustrates the kinetics of incorporation of $^3$H-dTMP into acid-insoluble product employing the purified particles isolated from the milk of a woman with a familial history of breast cancer. (b) Illustrates these kinetics using an aliquot of the same purified sample kept at 4°C for 48 hr. The standard 125 μl endogenous reaction mixture contained in micromoles: 8 of Tris HCl (pH 8.3), 1 of $MgCl_2$, 6 of KCl, 3 of dithiothreitol, 0.2 each of the unlabeled deoxyribonucleoside triphosphates (dATP, dGTP, dCTP), and 0.01 of dTTP. 25 μl of $^3$H-dTTP (2 mCi/ml) were added to give a final specific activity of 2000 cpm/pmole. The reaction contained 250 μg protein from concentrated pellets resuspended in TNE and 0.08% Nonidet P-40. The DNA polymerase reaction mixture utilizing dT:rA as a template contained the same concentrations of Tris HCl, $MgCl_2$, KCl, dithiothreitol, $^3$H-dTTP, Nonidet P-40, and protein as the endogenous reaction mixture. It also contained 8 μg/ml dT:rA, 0.04 μmoles of dATP and dTTP, giving a final specific activity of 513 cpm/pmole. All reactions were carried out at 37°C. At time of sampling, aliquots were withdrawn and precipitated with 1 ml of trichloroacetic acid-saturated sodium orthophosphate-saturated sodium pyrophosphate 1:1:1 in the presence of 60 μg of *E. coli* RNA as carrier. After 10 min the precipitable radioactivity was collected on a nitrocellulose filter and dried, and the radioactivity was determined in a liquid scintillation counter using BBOT scintillation fluid.

The picamoles incorporation of $^3$H-TMP into acid-insoluble product per 100 μg protein was recorded at the times indicated. The ribonuclease A (Worthington) used in all experiments was tested for the presence of contaminating DNase activity and was found negative.

The relationship of the human milk particle to known oncogenic RNA viruses and to human neoplasia is of obvious importance. It will now be possible to resolve by molecular hybridization the relationship between the human B-type particles and the similar agents associated with mammary tumors in mice and monkeys. Hybridization of the DNA synthesized by these type-B particles with the RNA isolated from malignant and benign human breast tumors, as well as to "normal" breast tissue, should also answer many questions concerning the relevance of these particles to human breast cancer.

## REFERENCES

1. Temin, H. M. (1964). *Natn. Cancer Inst. Monogr.* **17**, 557.
2. Cole, R. K. and Furth, J. (1941). *Cancer Res.* **1**, 951.
3. Temin, H. M. (1968). *Cancer Res.* **28**, 1835.
4. Temin, H. M. and Mizutani, S. (1970). *Nature, Lond.* **226**, 1211.
5. Baltimore, D. (1970). *Nature, Lond.* **226**, 1209.
6. Spiegelman, S., Burny, A., Das, M. R., Keydar, J., Schlom, J., Travnicek, M. and Watson, K. (1970) *Nature, Lond.* **227**, 563.
7. Schlom, J., Harter, D. H., Burny, A. and Spiegelman, S. (1971). *Proc. natn. Acad. Sci. U.S.A.* **68**, 182.
8. Schlom, J., Spiegelman, S. and Moore, D. (1971). *Nature, Lond.* **231**, 97.
9. Hall, B. D. and Spiegelman, S. (1961). *Proc. natn. Acad. Sci. U.S.A.* **47**, 114.
10. Sigurdsson, B., Palsson, P. A. and Grimsson, H. (1957). *J. Neuropath. exp. Neurol.* **16**, 389.
11. Sigurdsson, B. and Palsson, P. A. (1958). *Br. J. exp. Path.* **39**, 519.
12. Thormar, H. and Palsson, P. A. (1967). *In* "Perspectives in Virology" Vol. 5 (M. Pollard ed.), p. 291, Academic Press, New York and London.
13. Sigurdsson, B. (1954). *Br. vet. J.* **110**, 341.
14. Thormar, H. (1965). *In* "Slow, Latent and Temperate Virus Infections", Nat. Inst. Neurological Diseases and Blindness Monograph 2, p. 335. Washington.
15. Coward, J. E., Harter, D. H. and Morgan, C. (1970). *Virology* **40**, 1030.
16. Lin, F. H. and Thormar, F. H. (1970). *J. Virol.* **6**, 702.
17. Stone, L. B., Scolnick, E., Takemoto, K. and Aaronson, S. A. (1971). *Nature, Lond.* **229**, 257.
18. Harter, D. H., Schlom, J. and Spiegelman, S. (1971). *Biochim. biophys. Acta* **240**, 435.
19. Takemoto, K. I. and Stone, L. B. (1971). *J. Virol.* **7**, 770.
20. Spiegelman, S., Burny, A., Das, M. R., Keydar, J., Schlom, J., Travnicek, M. and Watson, K. (1970). *Nature, Lond.* **227**, 1029.
21. Kornberg, A. (1961). "Enzymatic Synthesis of DNA", p. 103, John Wiley, New York.
22. Goulian, M., Kornberg, A. and Sinsheimer, R. L. (1967). *Proc. natn. Acad. Sci. U.S.A.* **58**, 2321.
23. Denhardt, D. T. (1966). *Biochem. biophys. Res. Commun.* **23**, 641.
24. Gillespie, D. and Spiegelman, S. (1965). *J. molec. Biol.* **12**, 829.
25. Spiegelman, S., Burny, A., Das, M. R., Keydar, J., Schlom, J., Travnicek, M. and Watson, K. (1970). *Nature, Lond.* **228**, 430.

26. Lee-Huang, S. and Cavalieri, L. F. (1964). *Proc. natn. Acad. Sci. U.S.A.* **51**, 1022.
27. Kacian, D. L., Watson, K. F., Burny, A. and Spiegelman, S. (1971). *Biochim. biophys. Acta* **246**, 365.
28. Lehman, I. R., Bessman, M. S., Simms, E. S. and Kornberg, A. J. (1959). *J. biol. Chem.* **233**, 2733.
29. Bollum, F. J. (1959). *J. biol. Chem.* **234**, 2733.
30. Goodman, N. C. and Spiegelman, S. (1971). *Proc. natn. Acad. Sci. U.S.A.* **68**, 2203.
31. Chopra, H. C. and Mason, M. M. (1970). *Cancer Res.* **30**, 2081.
32. Jensen, E. M., Zelljadt, I. and Chopra, H. C. (1970). *Cancer Res.* **30**, 2388.
33. Ahmed, M. S., Mayyasi, S. A., Chopra, H. C., Zelljadt, I. and Jensen, E. M. (1971). *J. natn. Cancer Inst.* **46**, 1325.
34. Schlom, J. and Spiegelman, S. (1971). *Proc. natn. Acad. Sci. U.S.A.* **68**, 1613.
35. Moore, D. H., Charney, J., Kramarsky, B., Lasfargues, E. Y., Sarker, N. H., Brennan, M. J., Burrows, J. H., Sirsat, S. M., Paymaster, J. C. and Vaidya, A. B. (1971). *Nature, Lond.* **229**, 611.
36. Říman, J. and Beaudreau, G. S. (1970). *Nature, Lond.* **228**, 427.
37. Hatanaka, M., Huebner, R. J. and Gilden, R. V. (1970). *Proc. natn. Acad. Sci. U.S.A.* **67**, 143.
38. Scolnick, E. M., Aaronson, S. A. and Todaro, G. J. (1970). *Proc. natn. Acad. Sci. U.S.A.* **70**, 1789.
39. Green, M., Rokutanda, M., Fujinaga, K., Ray, R. K., Rokutanda, H. and Gurgo, C. (1970). *Proc. natn. Acad. Sci. U.S.A.* **67**, 385.
40. Gallo, R. C., Sarin, P. S., Allen, P. T., Newton, W. A., Priori, E. S., Bowen, J. M. and Dmochowski, L. (1971). *Nature, Lond.* **232**, 140.
41. Parks, W. P., Scolnick, E. M., Todaro, G. J. and Aaronson, S. A. (1971). *Nature, Lond.* **229**, 258.

## DISCUSSION

**O. P. van Diggelen:** Is Q$\beta$ RNA really used as a template for AMV-virus-enzyme or is there a DNA contamination in the preparation that serves as template?

**J. Schlom:** The reaction is sensitive to RNase. The DNA product of the reaction also hybridizes specifically to Q$\beta$ RNA.

**S. Danev:** Is the synthetic RNA-DNA template a specific activator of RNA-dependent DNA polymerase alone?

**J. Schlom:** The synthetic templates are not specific in all instances.

**S. Danev:** Is the activity of this enzyme elevated in all types of leukaemia?

**J. Schlom:** All types of leukaemias contain enzymes that respond well to dT:rA. This does not mean, however, that a reverse transcriptase is present.

**E. De Clercq:** Have you ever demonstrated an RNA-dependent DNA polymerase in EB or other DNA viruses?

**J. Schlom:** No, we have not looked for this.

**E. De Clercq:** Have you attempted to demonstrate the enzyme in cell cultures infected by mammary tumour viruses?

**J. Schlom:** No, but we plan to look for it.

**E. De Clercq:** Did I understand you to say that you did not find the enzyme in leukaemia cells? How do you explain this in view of other positive reports?

**J. Schlom:** Yes. We have examined numerous leukaemic white blood cell samples for the presence of reverse transcriptase and have not been able to detect this activity. This does not mean, however, that this enzyme is not present. Further studies must be pursued.

# Viral and Host Cell Interactions with 5-Iodo-2′-Deoxyuridine (Idoxuridine)*

## WILLIAM H. PRUSOFF

*Department of Pharmacology, Yale University,*
*New Haven, Connecticut, U.S.A.*

Although many compounds with antiviral activity are either synthesized or isolated from natural products, the number available for clinical trial are few. The majority of agents that exhibit antiviral activity in cell culture unfortunately are either inactive or too toxic under conditions *in vivo*. An active compound *in vitro* may be unsatisfactory *in vivo* for such number of reasons as metabolic inactivation, failure of transport into the cell, inadequate vascularity to the infected area, adverse effect on an essential cellular function, etc. Many a logically designed antiviral compound is buried in the graveyard of despair for one or more of these reasons.

Thus, for example, we studied the antiviral activity of 5,6-dichloro-1-(2′-deoxy-$\beta$-D-ribofuranosyl)benzimidazole (Fig. 1) and it exerted marked inhibition against the herpes simplex virus in cell culture [12]. Although *no* cytotoxicity could be observed visually, biochemical analysis of the effect of this compound's antiviral concentration on RNA, DNA, and protein synthesis by the uninfected host mammalian cell indicated that the biosynthesis of these polymers was inhibited by 55, 32, and 25% respectively [12]. Thus, the observed antiviral effects of this compound may be secondary to an effect on the metabolic processes of the host cell required for viral propagation. If the uninfected host cell *in vivo* can recover or be readily replaced in the tissue from such a biochemical insult, then the agent may have clinical utility. However, if the converse condition prevails, then the toxicity produced may be too high a price for the desired antiviral effect. Therefore, an understanding of the total impact of an antiviral substance on an infected host entails elucidation of the biochemical events that occur not only in the infected cell, but also in the uninfected tissue.

Another example of a failure is our observation that 6-azathymidine (Fig. 1) (an analogue of thymidine in which the carbon in position 6 of the pyrimidine moiety is replaced by a nitrogen) markedly inhibits the replication of herpes simplex virus in culture; however, thymidine in 0.015 molar equivalent completely prevented the inhibitory effect [50]. Thus, the probability of achieving the concentration of the analogue in the infected cells *in vivo* required to produce an antiviral effect would have a low probability; this was confirmed.

Recently, my colleague Dr. P. Chang synthesized the N-methyl analogue of 5-iodo-2'-deoxyuridine (Fig. 1) and my colleague Dr. B. Goz evaluated its effect on both the replication of T4td8 phage and herpes simplex virus. At a concentration of 5-iodo-2'-deoxyuridine (IdUrd) that inhibits replication of

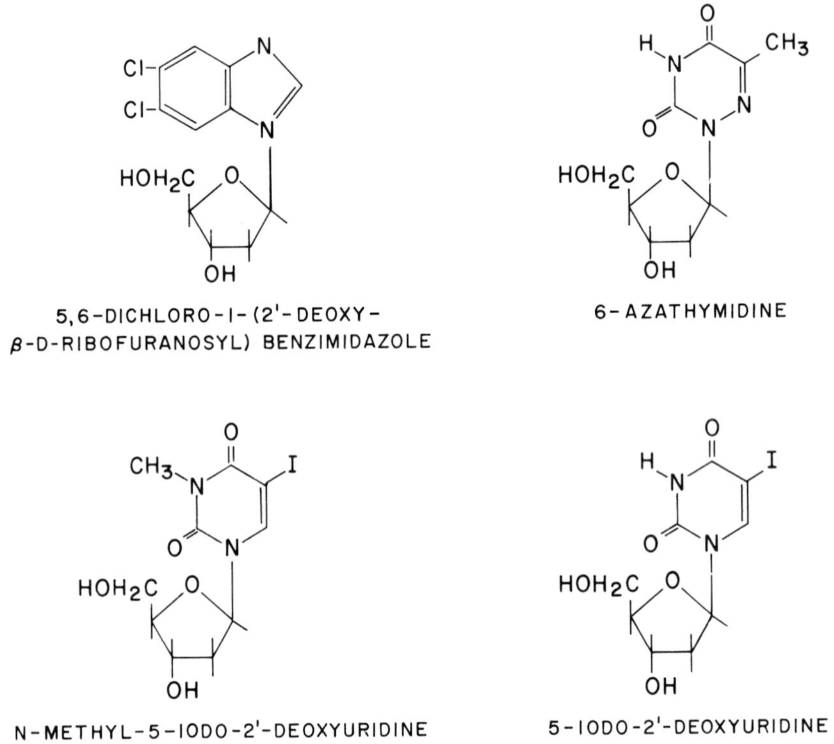

5,6-DICHLORO-1-(2'-DEOXY-β-D-RIBOFURANOSYL) BENZIMIDAZOLE

6-AZATHYMIDINE

N-METHYL-5-IODO-2'-DEOXYURIDINE

5-IODO-2'-DEOXYURIDINE

Figure 1. Structure of several deoxyribonucleosides with antiviral activity.

T4td8 phage, the N-methyl analogue is inert. Similar studies with herpes simplex virus showed that a 100-fold increased concentration of the N-methyl analogue is required to achieve the same degree of inhibition as that produced by IdUrd. Kinetic studies with a highly purified preparation of *E. coli* thymidine kinase, performed by my colleague Dr. P. Voytek [55], revealed that the N-methyl analogue of IdUrd is an uncompetitive inhibitor with respect to thymidine, whereas the parent analogue, IdUrd (in agreement with previous findings) is a competitive inhibitor. Competitive inhibition kinetics were found, however with respect to ATP-$Mg^{2+}$. The potential role of this compound as an antiviral agent is under study, although we are concerned about its inhibition of the ATP-$Mg^{2+}$ function, which, if similarly affected in other kinases, may prove quite toxic.

The remainder of this discussion is restricted primarily to cellular and antiviral activities of 5-iodo-2′-deoxyuridine (Idoxuridine, IdUrd, IUDR, IDU) (Fig. 1). This analogue of thymidine is in current clinical use for the therapy of herpes simplex keratitis, a major cause of blindness in man. Recent reviews of the biochemical and chemotherapeutic aspect of this drug in relation to other antiviral agents have appeared [5, 23, 27, 48].

A discussion of the effect of IdUrd on the uninfected cell or host is pertinent because this compound is not the ideal antiviral agent since its action is not restricted to a virus-specified biochemical reaction. Nevertheless, it appears that IdUrd does indeed exert its antiviral effect at a concentration that does not produce cell toxicity. Cramer et al. [11] found inhibition of herpes simplex virus at a concentration 0.1 of that required to inhibit reproduction of the host cell. Hanna [25] similarly observed that IdUrd prevented the formation of infectious herpes simplex virus in rabbit cornea in concentrations that did *not* affect the synthesis of DNA by the normal basal epithelium cells of the cornea.

What effect does IdUrd have on the uninfected cell? Three major areas have been observed where IdUrd or the appropriate phosphorylated derivatives exert marked inhibitory effects: (a) competitive inhibition of several enzymes concerned with the biosynthesis of DNA-thymine (thymidine kinase, thymidylate kinase and DNA polymerase) due to the analogue or the appropriate phosphorylated derivative functioning as an alternate substrate; (b) allosteric or feedback inhibition by the triphosphate of IdUrd, mimicking the normal regulatory activity of deoxythymidine triphosphate on thymidine kinase, deoxycytidylate deaminase, and cytidine diphosphate reductase; (c) incorporation into DNA with subsequent effect in expression of genetic information either during replication or transcription.

One or more of these biochemical parameters causes clinical toxicity in man. The toxicity is dose-dependent and has been observed in cancer patients systemically receiving approximately 100 mg/kg of IdUrd for four or more days. This toxicity is manifest as leucopenia, thrombocytopenia, alopecia, stomatitis, or transverse ridging of the nails. These considerations are important in the administration of IdUrd during therapy of such systemic viral infections as viral encephalitis, but are of no consequence in IdUrd topical therapy of herpes simplex keratitis in man. In the latter situation, the dose of IdUrd instilled into the conjunctival sac, even if totally absorbed systemically, is equivalent to 0.001 the amount required to produce any of these toxicities. Furthermore, any IdUrd absorbed systemically would be subject to rapid catabolism, primarily by the liver, sequentially forming 5-iodouracil (+ deoxyribose-1-phosphate), uracil (+ iodide), dihydrouracil, and finally acyclic catabolic derivatives of dihydrouracil. In fact, one major problem in the systemic use of IdUrd is its extremely rapid metabolic degradation, which undoubtedly limits its incorporation into DNA (Fig. 2). Barton and Tobin [3] observed that systemic administration of IdUrd to children for treatment of viral infections did not produce adverse toxic reactions. This was confirmed in similar studies by two other investigators. However, other reports have indicated that toxicity does occur in this age category.

The potential role of IdUrd in neoplasm therapy in man, for which it was originally prepared [38], is still under investigation, particularly in combination with radiation. Meanwhile, it has found a major role as an antiviral agent. The antiviral potential of IdUrd was first observed by Herrmann [26], and used in the therapy of herpes simplex infection of eye in rabbits and man by Kaufman [31]. Although the replication of DNA-viruses (e.g., herpes simplex, vaccinia, pseudorabies, adenovirus, and polyoma) is inhibited by IdUrd, RNA-viruses are not affected, with the exception of Columbia-SK encephalitis and Rous sarcoma virus. The Rous sarcoma virus is inhibited only when IdUrd is administered to

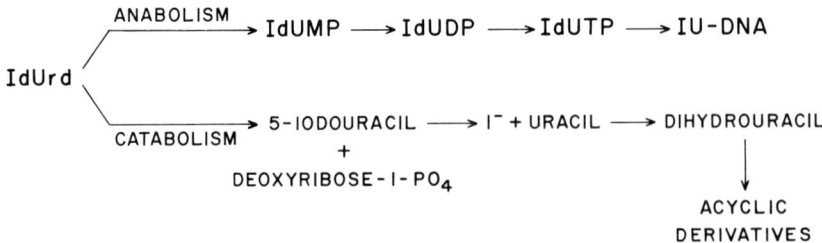

Figure 2. Metabolism of 5-iodo-2'-deoxyuridine.

cells at the time of infection [17], and may be related to the discovery by Temin [54] and Baltimore [2] that RNA tumor viruses, in their process of replication, act as a template for the synthesis of DNA. It would be of much interest to determine whether IdUrd does incorporate into DNA and thereby prevent cell transformation. The induction of tumors by the DNA viruses, adenovirus type 12, and polyoma virus in newborn hamsters has been markedly reduced by the administration of IdUrd [16, 28].

The efficacy of IdUrd in the topical therapy of herpes simplex infection of the corneal epithelium in man has been confirmed; however, its effectiveness in deep stromal infections is doubtful. Such nucleosides as trifluorothymidine, cytosine arabinoside, and adenine arabinoside are under experimental study for therapy of herpetic keratitis and other viral diseases.

Cutaneous herpes infections have been treated with IdUrd. The results of such topical therapy are controversial—the problem being one of insuring penetration of the drug into the infected cells. MacCallum and Juel-Jensen [34] successfully treated herpes simplex lesions of the skin in man by application of a

5% solution of IdUrd in dimethylsulfoxide. This solvent provides about a ten-fold greater solubility of the drug relative to water and may also result in more effective penetration of the drug into cells.

Herpes varicella virus is the etiological agent responsible for varicella (chicken pox) and herpes zoster (shingles). Although the replication of herpes varicella virus and the zoster virus were markedly affected by IdUrd in culture, therapy of herpes zoster by topical application in man was not effective [35]. The suggestion was made that since the virus resides in the posterior root ganglia, the drug may not have reached the site of viral replication. Juel-Jensen et al. [29], however, using a very concentrated solution of IdUrd (40% in dimethyl sulfoxide), reported accelerated healing and a reduction in the duration of pain from an average of over a month to that of less than three days. Waltuch and Sachs [56] treated disseminated herpes zoster, which developed in a patient with Hodgkin's disease, with systemically administered IdUrd and reported an excellent response.

Herpes genitalis has been treated with a 0.5% ointment and beneficial effects have been reported.

Conchie et al. [10] reported cytomegalovirus to be favorably affected by IdUrd in that the urinary excretion of the virus was reduced; however, further study [3] indicated that clinical improvement had not been achieved.

Success in the treatment of various localized infections with IdUrd is based on the ability to achieve an adequate concentration in a restricted area. The logical extension is the determination of whether systemic infections could also be suppressed. The extensive experience of the Yale group [6, 58] in the treatment of neoplasms in man by systemic administration of IdUrd led to the experiments of Calabresi et al. [7], who showed that vaccinia infections, both in rabbits and in patients with advanced neoplastic disease, could be suppressed by the intravenous administration of well-tolerated amounts of IdUrd. Although the acute toxicity (stomatitis, alopecia, hematopoietic depression) observed was reversible [7], it has been recommended that the use of this compound be restricted to patients with severe or potentially lethal DNA-viral infections because incorporation of IdUrd into the genome of the host cell conceivably could express itself by genetic damage, infertility, or even neoplastic change [57]. Although FdUrd, BdUrd and IdUrd inhibit antibody formation *in vitro*, BdUrd, not IdUrd, inhibited the immune response in animal systems (literature cited in [39]). This may be an advantage in IdUrd use for systemic antiviral therapy because a possible host defense mechanism is apparently not impaired by this drug. The teratogenic potential of these halogenated deoxyribonucleosides, including IdUrd, has also been discussed in reference [39].

An exciting use of IdUrd is the elucidation of its potential role in the systemic therapy of herpes encephalitis. The initial report by Breeden et al. [4] described the successful treatment of a patient with encephalitis caused by herpes simplex virus by administering an intravenous injection of IdUrd (total dose 39 g) during a seven-day period after surgical decompression. This provocative finding has stimulated numerous investigators. However, only a few case reports have been presented. Although not uniformly enthusiastic, reports clearly indicate the need for a properly controlled study of IdUrd in the treatment of herpetic encephalitis. It is obviously essential to establish that the disease entity under study is caused by a herpes virus or other DNA virus. Some investigators have assumed that this disease is uniformly fatal or disabling; but fortunately for the patient, this is not so. Thus, interpretations of reports that describe treatment of relatively few cases are subject to valid questions. Nevertheless, the majority of these reports indicate that IdUrd merits continued evaluation in the therapy of this dread disease. Of particular importance is the study of Rappel and Brihaye [44], who observed favorable response in 9 of 11 cases treated with IdUrd.

A number of other purine and pyrimidine derivatives have demonstrated marked antiviral activity *in vivo*, and these have been tabulated by Schabel and Montgomery [48]. Several have been reviewed recently [23].

Cytosine arabinoside is not only an effective drug against certain neoplasms, but also has demonstrated marked antiviral activity. Although effective in herpes keratitis, it has been reported to be toxic to the cornea, producing ulceration of the corneal epithelium. Cytosine arabinoside has been used in the treatment of disseminated varcilla-zoster (shingles), herpes simplex keratitis, as well as in several cases of herpes encephalitis (literature cited in [48]). Although the potential role in therapy is not established, results are encouraging.

Adenine arabinoside has a broad spectrum of activity against DNA viruses in both culture and animal systems. Although massive doses were required (3 g per kilogram), this compound is effective against both herpes simplex and vaccinia encephalitis in experimental animals. Marked activity has been reported in therapy of stromal involvement of herpes keratitis. No toxicity has been reported and clinical trials are presumably in progress (literature cited in [48]).

6-Azauridine at relatively high concentrations has been shown to inhibit certain RNA and DNA viruses [15, 43]. Preliminary clinical trials of pateints with herpes infection of the eye were encouraging [37]. A good therapeutic response in the treatment of smallpox in man with 6-azauridine has been reported recently.

5-Ethyl-2'-deoxyuridine was synthesized by Gauri [18] and Swierkowski and Shugar [52], as well as the radioactive compound, Gauri *et al.* [20]. Activity against herpes simplex virus, vaccinia virus, and Aujeszky virus has been reported [19, 52]. More details of this very interesting compound are presented by Dr. Shugar in this symposium.

5-Trifluoromethyl-2'-deoxyuridine was synthesized by Heidelberger of the University of Wisconsin and found to be a very potent antiviral agent, particularly effective against herpes keratitis.

Whereas the aim in the chemotherapy of neoplasms is the total destruction of the neoplastic cell, it is not essential to have complete eradication of the virus to achieve successful antiviral therapy. For example, it has been reported that if a patient with smallpox has a virus blood level of $10^4$ infecting doses (chorioallantoic membrane assay) or higher in the second day of illness, the patient will probably die; however, if the titer were $10^2$ infecting doses per ml or less, the viremia will disappear and the patient recovers. Thus, the body is able to recover from at least certain virus infections by utilization of host defense mechanisms, provided the intensity of the infection is not above a critical level. Such drugs as those discussed above are capable of achieving these reductions in virus concentration.

How does IdUrd (Fig. 1) exert its antiviral effect? The replacement of the methyl group in position 5 of thymidine with a halogen alters the electron configuration of the pyrimidine moiety because of the inductive effect of the halogen; this results in more acidic dissociation constants. The pKa of thymidine is 9.8, whereas that of IdUrd is 8.25. Because the N-1 position is substituted with the deoxyribose moiety, the proton on the N-3 position must be dissociated. Steric hindrance because of the replacement of the methyl group in thymidine by either a bromine, chlorine, or iodine atom does not constitute a problem because their van der Walls radii ($CH_3$, 2.00; chlorine, 1.80; bromine, 1.95; iodine, 2.15) are reasonably similar. Therefore, it is not surprising that these halogenated deoxyribonucleosides and their phosphorylated derivatives participate in reactions that thymidine and its phosphorylated derivatives enter.

In an attempt to understand the antiviral activity of IdUrd, we initially asked, "Are the enzymes concerned with the formation of the viral components, specifically viral-DNA, in the virus-infected cells more sensitive to inhibition by the indicated analogue or its phosphorylated derivatives than are the corresponding enzymes of the infected cells?" [41]. The expectation that a difference might be found in susceptibility to inhibition, and thus selective toxicity, was based on many observations of increased activities of enzymes that form precursors of DNA-thymidine and DNA, which occur upon infection of cells with many viruses. In many instances, the virus "induced" enzymes have physicochemical properties that differ from the corresponding enzyme present in uninfected cells. Therefore, our approach has been to identify the enzyme concerned with the formation of virus-DNA that is primarily inhibited, and to determine the consequences of incorporation of IdUrd into the DNA of the virus. We found *no* significant differences in susceptibility to inhibition by IdUrd or its monophosphate derivative in the thymidine kinase and thymidylate kinase of mammalian cells (African green monkey kidney cells, BSC-1) before and after infection with herpes simplex virus [41].

The triphosphate of IdUrd was prepared by my colleague Dr. P. K. Chang [42], and it effectively replaces, on a molar basis, the utilization of radioactive dTTP for the biosynthesis of DNA by a particulate-free extract of a murine neoplasm [1]. Thus, the utilization of dATP for the formation of DNA was not depressed even by a four-fold excess of the triphosphate of IdUrd over that required to produce 50% inhibition of the utilization of dTTP. In agreement with findings of Russell *et al.* [47], infection of cells in culture with herpes simplex virus results in a progressive increase in DNA-polymerase activity [50]. No differences were observed in the susceptibility to inhibition by IdUTP of DNA polymerase derived from uninfected cells and from cells infected with herpes simplex virus [50].

As indicated above, IdUTP mimics dTTP in exerting an allosteric or feedback inhibition of thymidine kinase, deoxycytidylate deaminase, and cytidine diphosphate reductase. No difference was observed in susceptibility to inhibition of thymidine kinase by dTTP in non-infected cells and cells infected with adenovirus [33] or herpes simplex virus [13]. However, thymidine kinase, specified by herpes simplex virus [32] and by vaccinia virus [24], was reported to be significantly *less* susceptible to inhibition by dTTP than the enzyme present in the uninfected host cell. A comparison was made of the relative susceptibility of deoxycytidylate deaminase to inhibition by IdUTP and by dTTP. Levels of the enzyme present in uninfected cells and cells infected with herpes simplex virus showed that although the amount of enzymic activity increased almost three-fold during infection, there was no alteration in the susceptibility of the "induced" enzyme by either of the triphosphate derivatives [14].

The inhibitions described above are of the competitive type and thus readily reversible either by increase in the normal substrate or by metabolic conversion of the appropriate analogue. Thus, we believe that the primary site of inhibition by IdUrd is in an event that is subsequent to incorporation of the antimetabolite into the viral DNA in substitution of the thymidine moiety. A number of very important biochemical and physical effects result from such incorporation of halogenated deoxyribonucleosides into DNA (literature cited in [23, 39], which include: (a) increased rate of mutation, (b) increased number of errors in protein formation, (c) inhibition of cellular reproduction, and (d) increased sensitivity to X-, ultraviolet, and near visible radiations.

Both BdUrd and IdUrd have been demonstrated to be incorporated into the DNA of mammalian cells, bacteria, phage, and animal viruses. These halogenated deoxyribonucleosides have no direct effect on either extracellular vaccinia or herpes simplex viruses, and no effect on the adsorption of herpes virus to cells. However, they markedly inhibit the formation of infectious virus particles because the ratio of virus particles to infectious virus may be as high as $10^7:1$ (literature cited in [23, 39]). The virus particles formed in the presence of IdUrd and BdUrd are abnormal in appearance (literature cited in [39]).

The effect of IdUrd and BdUrd varies with the specific virus under study. Kaplan and Ben-Porat [30] observed that pseudorabies virus grown in the presence of BdUrd produced non-infectious yet complete virus particles, whereas

IdUrd permitted substituted DNA to be synthesized, which could not be incorporated into a virus particle unless thymidine was added either before or after the formation of the halogenated DNA. Prusoff *et al.* [40] found IdUrd was incorporated into complete vaccinia virus particles to the extent of 18% replacement of viral DNA-thymidine. McCrea and Lipman [36] examined the DNA in normal and IdUrd-labeled DNA derived from this vaccinia virus by a modification of the Kleinschmidt method, and found the IdUrd-substituted DNA liberated by osmotic shock extensively fragmented, in contrast to the normal DNA, which was consistently relatively long and continuous. These investigators proposed that the decreased pock-forming ability of the IdUrd-substituted virus particles may be attributed, in part, to the increased fragility of the DNA when released within the host cell by the "uncoating enzyme".

Recent studies by my colleague, Dr. B. Goz [21, 22], have been concerned with the ability of phage-containing IdUrd-substituted DNA to perform the normal biochemical steps necessary for replication in the absence of IdUrd in the growth medium. IdUrd-substituted T4 phage were unable to induce normal levels of several enzymes: dCMP hydroxymethylase, lysozyme, dihydrofolate reductase, and thymidylate synthetase [21]. Phenotypic and genotypic rescue experiments [22] indicated that IdUrd-substitution for thymidine in the genome of phage T4 causes a differential inactivation that may well be related to the differential substitution of IdUrd for thymidine in the various genes. That some genes are richer in thymidine residues is supported experimentally by Szybalski *et al.* [53], who demonstrated thymidine-rich clusters in T2 DNA.

Recent studies by Stillwagen and Tomkins [51] in a mammalian cell system *in vitro* show that BdUrd also causes a selective inhibitory effect of certain genes, as evidenced by a decreased formation of specific enzymes (tyrosine aminotransferase, lactate dehydrogenase, alcohol dehydrogenase, and glucose-6-phosphate dehydrogenase). Such enzymes as malate dehydrogenase, acid phosphatase, or alanine aminotransferase were *not* significantly affected. These authors postulated that BdUrd exerts its effect by slowing the transcription rate of certain messenger RNAs.

A possible physicochemical basis for the findings of Stellwagen and Tomkins [51] and Goz and Prusoff [21, 22] may have been presented by the studies of Camerman and Trotter [9], who examined the crystal and molecular structure of IdUrd by X-ray diffraction and found an unusually short intermolecular distance of 2.96 Å. They postulated that the ability of iodine to form charge transfer bonds may cause an augmented interchain attraction that could either prevent or delay synthesis of virus-DNA. An extension of the hypothesis of Camerman and Trotter [9] that may apply to the observations of Stellwagen and Tomkins [51] and Goz and Prusoff [21, 22] is that there may be a delay in DNA transcription due to the formation of the postulated charge transfer complex.

The increased rate of mutation, as well as increase in errors of protein formation, that result from incorporation of the halogenated uracil derivatives in the nucleic acid polymer has been explained classically by the increased probability of base pair errors caused by the lower pKa of the halogenated compounds. Thus, relative to thymidine, IdUrd at physiological pH exists with a 34-fold greater probability in the anionic form. This has been interpreted to mean that a greater probability of base pairing with a guanine derivative exists. Thus, during replication, an A-T pair is converted into a G-C pair, and during transcription the messenger RNA may have a guanine inserted instead of an adenine. The adverse consequence of such alterations could be catastrophic.

Finally, we discuss another property of the halogenated uracil derivatives, their ability to sensitize mammalian cells, bacteria, phage and viruses to the lethal effect of X- and ultraviolet radiations subsequent to incorporation into the nucleic acid polymer. My former colleague, Dr. Rupp, performed studies of the photochemistry of 5-iodouracil in an attempt to elucidate the photochemical basis for this phenomenon [45, 46]. The initial photochemical reaction is the formation of a uracilyl free radical by a dehalogenation mechanism. The formed free radical is capable not only of abstracting a hydrogen from an appropriate organic donor to form uracil, but also of cleaving a disulfide with formation of uracil-5-thioether [45, 46].

A study concerned with the possible sensitization of enzymes to inactivation by radiation when the appropriate iodinated substrate is present at the active site of the enzyme was initiated. Appropriate enzyme-substrate complexes considered for study include: thymidine kinase-IdUrd, thymidylate kinase-IdUMP, DNA polymerase-IdUTP and deoxycytidylate kinase-IdCMP.

Irradiation of thymidine kinase with uv-radiation (253.7 nm) resulted in characteristic first-order inactivation kinetics. Thymidine reduces, and IdUrd enhances, the rate of inactivation of thymidine kinase [8]. Evidence has been obtained that both effects are active site directed, thus highly specific. The $K_m$ for thymidine equalled the concentration of thymidine required to produce 50% protection. However, the concentration required to produce 50% inactivation by IdUrd was two to three times that of its $K_m$ for this enzyme. A possible explanation is the formation of deoxyuridine from IdUrd at the active site by abstraction of a hydrogen from either the buffer or a non-sensitive amino acid rather than the formation of an inactivating complex. The involvement of an active site directed process is supported by the following observations: (a) allosteric regulators affect thymidine protection and IdUrd inactivation in accordance with their effect on the $K_m$ of these substrates, (b) the site affected by thymidine and IdUrd can be saturated, (c) the IdUrd sensitizing effect of thymidine kinase can be prevented by increasing amounts of thymidine. The photochemistry of IdUrd was investigated; like 5-iodouracil, the initial reaction is a dehalogenation with the formation of a deoxyuridine free radical. 5-Bromo-,

5-chloro- and 5-fluoro-2′-deoxyuridine do not sensitize thymidine kinase to uv-inactivation [55]. Under the conditions of irradiation, the bromo- and chloro-derivatives are photochemically inert, whereas the fluoro-derivatives form a hydrate rather than undergo dehalogenation.

The $N$-methyl derivative of IdUrd, previously described as having 100-fold lower potency against herpes simplex virus, has been shown by my colleague Dr. Voytek to be similarly less potent than IdUrd in the sensitization of thymidine kinase to uv inactivation. Dr. Voytek found the $N$-methyl analogue to be an uncompetitive inhibitor with respect to thymidine and a competitive inhibitor with respect to Mg-ATP. However, both substrates (ATP-Mg and thymidine) are required to protect against uv-sensitization of thymidine kinase by the $N$-methyl analogue.

Since irradiation of thymidine kinase in the presence of $^{14}$C-IdUrd, but not $^{125}$I-IdUrd, forms a radioactive covalent complex [8], we now have the opportunity to elucidate and compare the amino acid sequence at the active site of thymidine kinase, whose synthesis is directed by the host cell relative to the enzyme specified by the viral genome.

## ACKNOWLEDGEMENT

This research was supported by United States Public Health Service Research Grant CA-05262.

## REFERENCES

1. Bakhle, Y. S. and Prusoff, W. H. (1969). *Biochem. biophys. Acta* **174**, 302.
2. Baltimore, D. (1970). *Nature, Lond.* **226**, 1209.
3. Barton, B. W. and Tobin, J. O'H. (1970). *Ann. N.Y. Acad. Sci.* **173**, 90.
4. Breeden, C. J., Hall, T. C. and Tyler, H. R. (1966). *Ann. Intern. Med.* **65**, 1050.
5. Brown, D. C. (1971). *J. Ophthalm. Otolar.* **10**, 210.
6. Calabresi, P., Cardoso, S. S., Finch, S. C., Kligerman, M. M., Von Essen, C. F., Chu, M. Y. and Welch, A. D. (1961). *Cancer Res.* **21**, 550.
7. Calabresi, P., McCollum, R. W. and Welch, A. D. (1963). *Nature, Lond.* **197**, 767.
8. Cysyk, R. and Prusoff, W. H. *J. biol. Chem.* (1972). In press.
9. Camerman, N. and Trotter, J. (1964). *Science, N.Y.* **144**, 1348.
10. Conchie, A. F., Banton, B. W. and Tobin, J. O'H. (1968). *Br. med. J.* **4**, 162.
11. Cramer, J. W., Wacker, A. and Welch, A. D. (1963). *Biochem. J.* **87**, 26P.
12. Diwan, A., Gowdy, C. N., Robins, R. K. and Prusoff, W. H. (1968). *J. gen. Virol.* **3**, 393.
13. Diwan, A., Goz, B. and Prusoff, W. H. Unpublished experiment.
14. Diwan, A. and Prusoff, W. H. (1968). *Virology* **34**, 184.

15. Falke, D. and Rada, B. (1970). *Acta virol.* **14**, 115.
16. Fischer, D. S., Black, F. L. and Welch, A. D. (1965). *Nature, Lond.* **206**, 839.
17. Force, E. E. and Stewart, R. C. (1964). *Proc. Soc. exp. Biol. Med.* **116**, 803.
18. Gauri, K. K. (1969). *British Patent* **1**, 170, 565.
19. Gauri, K. K., Malorny, G. and Riehm, E. (1970). *Albrecht v. Graefes Arch. Klin. exp. Ophthal.* **179**, 287.
20. Gauri, K. K., Pflughaupt, K. W. and Müller, R. (1969). *Z. Naturf.* **246**, 833.
21. Goz, B. and Prusoff, W. H. (1968). *J. biol. Chem.* **243**, 4750.
22. Goz, B. and Prusoff, W. H. (1970). *Ann. N.Y. Acad. Sci.* **173**, 379.
23. Goz, B. and Prusoff, W. H. (1970). *A. Rev. Pharmacol.* **10**, 143.
24. Green, M., Pinna, M. and Chagoya, V. (1964). *J. biol. Chem.* **239**, 1188.
25. Hanna, C. (1968). *Exp. Eye Res.* **5**, 164.
26. Herrmann, E. C., Jr. (1961). *Proc. Soc. exp. Biol. Med.* **107**, 142.
27. Herrmann, E. C. Jr. and Stineberg, W. R. (Eds). (1970). "Second Conference on Antiviral Substances", *Ann. N.Y. Acad. Sci.* **173**.
28. Heubner, R. J., Lane, W. T., Welch, A. D., Calabresi, P., McCollum, R. W. and Prusoff, W. H. (1963). *Science, N.Y.* **142**, 488.
29. Juel-Jensen, B. E., MacCallum, F. O., Mackenzie, A. M. R. and Pike, M. C. (1970). *Br. med. J.* **4**, 776.
30. Kaplan, A. S. and Ben-Porat, T. (1966). *J. molec. Biol.* **19**, 320.
31. Kaufman, H. E. (1965). *Prog. med. Virol.* **7**, 116.
32. Klemperer, H. G., Haynes, G. R., Shedden, W. I. H. and Watsen, D. K. (1967). *Virology* **31**, 120.
33. Ledinko, N. (1967). *Cancer Res.* **27**, 1459.
34. MacCallum, F. O. and Juel-Jensen, B. E. (1966). *Br. med. J.* **2**, 805.
35. McCallum, D. I., Johnston, E. N. M. and Raju, B. H. (1964). *Br. J. Derm.* **76**, 459.
36. McCrea, J. F. and Lipman, M. B. (1967). *J. Virol.* **1**, 1037.
37. Myska, V., Elis, J., Plevova, J. and Raskova, H. (1967). *Lancet* **1230**.
38. Prusoff, W. H. (1959). *Biochim. biophys. Acta* **32**, 295.
39. Prusoff, W. H. (1967). *Pharmac. Rev.* **19**, 209.
40. Prusoff, W. H., Bakhle, Y. S. and McCrea, J. F. (1963). *Nature, Lond.* **199**, 1310.
41. Prusoff, A. H., Bakhle, Y. S. and Sekely, L. (1965). *Ann. N.Y. Acad. Sci.* **130**, 135.
42. Prusoff, W. H. and Chang, P. K. (1968). *J. biol. Chem.* **243**, 223.
43. Rada, B. and Blaskovic, D. (1961). *Acta virol.* **5**, 308.
44. Rappel, M. and Brihaye, J. (1969). *Revue Neurol.* **121**, 93.
45. Rupp, W. D. and Prusoff, W. H. (1964). *Nature, Lond.* **202**, 1288.
46. Rupp, W. D. and Prusoff, W. H. (1965). *Biochem. biophys. Res. Commun.* **18**, 145, 158.
47. Russell, W. C., Gold, E., Keir, H. M., Omura, H., Wilson, D. H. and Wildy, P. (1964). *Virology* **22**, 103.
48. Schabel, F. M. and Montgomery, J. A., *In* "The International Encyclopedia of Pharmacology and Therapeutics" (D. J. Bauer ed.), Section 61. (In press).
49. Sekely, L. and Prusoff, W. H. (1966). *Nature, Lond.* **211**, 1260.
50. Sekely, L. and Prusoff, W. H. (1969). *Pharm. Res. Commun.* **1**, 3.
51. Stellwagen, R. H. and Tomkinks, G. M. (1971). *Proc. natn. Acad. Sci. U.S.A.* **68**, 1147.

52. Swierkowski, M. and Shugar, D. (1969). *J. med. Chem.* **12**, 533.
53. Szybalski, W. H., Kubinski, H. and Sheldrick, P. (1966). *Cold Spr. Harb. Symp. Quant. Biol.* **31**, 123.
54. Temin, H. M. and Mizutami, S. (1970). *Nature, Lond.* **226**, 1211.
55. Voytek, P., Chang, P. K. and Prusoff, W. H. (1972). *J. biol. Chem.* **247**, 367.
56. Waltuck, G. and Sachs, F. (1968). *Archs. intern. Med.* **121**, 458.
57. Welch, A. D. (1965). *Ann. N.Y. Acad. Sci.* **123**, 19.
58. Welch, A. D. and Prusoff, W. H. (1960). *Cancer Chemother. Rep.* **6**, 29.

## DISCUSSION

**P. Langen:** According to Szybalski, BUdR sensitizes cells to UV radiation to a greater extent than FUdR, while the reverse appears to hold for your enzyme system. Could you comment on this discrepancy?

**W. H. Prusoff:** Photochemical studies on IUdR and BUdR unequivocally show that the former in aqueous medium at neutral pH is photolabile, whereas BUdR is not. This presumably is related to the difference in strengths of the C–I bond as compared to the C–Br bond. In DNA, energy absorbed by other purine or pyrimidine components is probably transferred in sufficient amount to the C–Br bond of BUdR residues to cause dehalogenation.

**E. De Clercq:** In view of the breakdown of vaccinia virus DNA which has been obtained from Idoxuridine-treated cells, can you comment on (or have you looked at) the interferon-inducing capacity of vaccinia virus in the presence of idoxuridine? Such an approach might prove useful in elucidation of the mechanism of interferon production by DNA viruses.

**W. H. Prusoff:** We have not ourselves investigated this parameter. There is no evidence that degradation of halogenated DNA occurs in cells *per se.* The method of isolation, although mild, may still exert sufficient stress to degrade the DNA. Studies with other viruses in the presence of IUdR indicated that the production of interferon may be increased, decreased or remain unchanged relative to that produced in the absence of this analogue.

**D. Shugar:** Have you measured the extent of incorporation of IUdR into the DNA of vaccinia virus?

**W. H. Prusoff:** Yes. In collaboration with Dr. Bakhle and Dr. McCrea it was shown that about 18% of the DNA thymine was replaced by IUdR.

**H.-E. Jacob:** To what extent is the iodouracil, resulting from cleavage of IUdR, metabolized by the cells? And do you observe the same effects?

**W. H. Prusoff:** Following formation of iodouracil by enzymic cleavage of IUdR in the cell, the iodouracil itself possesses no antiviral activity. Under special conditions where purine nucleosides are provided to the cell to supply a pool of deoxyribose phosphate, iodouracil has been shown to be utilized.

**D. Shugar:** In view of the low pK of IUdR (about 8.0), can you tell me whether, at physiological pH, it is the neutral or monoanionic form of IUdR which is the active species? The potential antimetabolic role of the monoanionic forms of the 5-halogenouracils and their glycosides (all of which possess pK values in the neighbourhood of 8) appears to have been generally disregarded or merely overlooked.

**W. H. Prusoff:** It is difficult to answer this question since varying the solution pH may not affect the intracellular pH or, if it does, then other effects may intervene. The effects of halogenated uracil derivatives on specific enzymes have been studied, and the most striking effect is an inhibition of *E. coli* thymidine kinase by IdUTP at low pH and activation at high pH.

**D. Shugar:** Did I correctly understand you to say that, following cleavage of the glycosidic bond of IUdR, the resulting 5-iodouracil undergoes enzymic dehalogenation? If so, how?

**W. H. Prusoff:** This is an enzymatic process extensively studied by Dr. Grisola many years ago, and subsequently in our own laboratories in connection with the biological properties of 5-iodocytosine (see Y. S. Bakhle, W. A. Creasy, A. C. Sartorelli and W. H. Prusoff, *Biochem. Pharmac.* **13**, 1249, 1946). The process may be only partially enzymic. Dihydrouracil dehydrogenase is the enzyme responsible for the reduction of 5-iodouracil to the 5,6-dihydrouracil derivative which is then believed to spontaneously lose HI with the formation of uracil.

# Antiviral Action of Oxidized Polyamines

U. BACHRACH

*Institute of Microbiology,*
*Hebrew University-Hadassah Medical School,*
*Jerusalem, Israel*

## INTRODUCTION

The polyamines spermine, $NH_2(CH_2)_3NH(CH_2)_4NH(CH_2)_3NH_2$ and spermidine, $NH_2(CH_2)_3NH(CH_2)_4NH_2$, are ubiquitous polycations which tend to form complexes with polyanions, including nucleic acids [cf. 1,2]. The specific conformation of the polyamine molecule facilitates a highly stereospecific interaction with double-stranded polynucleotides [3, 4] and may thus stabilize molecules, like DNA, against thermal denaturation and enzymic degradation. It is conceivable that a substituted polyamine, in which amino groups are replaced by more reactive radicals, would even prevent strand separation, and thus interfere with the expression of the genetic material. In other words, it is to be expected that such a substituted polyamine would behave as an antimicrobial agent.

## ANTIMICROBIAL ACTION

During the last several years we explored this possibility and used oxidized polyamines as substituted polyamine molecules [cf. 5]. These compounds, which are obtained by the enzymic oxidation of either spermine or spermidine, are aminoaldehydes [6] (Fig. 1). Numerous studies indicated that these primary oxidation products may undergo secondary reactions (such as condensation [7] and β-elimination [7-9]) mainly under alkaline conditions and at elevated temperatures [7].

Previous studies in our laboratory indicated that the growth of various bacteria (gram-positive, gram-negative and Mycobacteria) was inhibited by oxidized spermine at a concentration of 20 μg/ml [10]. We also demonstrated that oxidized spermine inactivated coliphages of the T-odd series [11-12], while T-even phages were relatively resistant (Table 1). These findings were confirmed in other laboratories [13-15]. Oxidized spermine also inactivates some plant viruses [16], including tobacco mosaic virus (TMV) and various animal viruses [17, 18]. In all these experiments viruses were incubated with the drug for several hours and then assayed for the infectivity (Table 2).

Table 1. Effect of oxidized spermine on the viability of various bacteriophages

| Phage | Plaque-Forming Units | |
|---|---|---|
| | Control | Oxidized Spermine |
| T1 | $1.8 \times 10^9$ | $1.2 \times 10^5$ |
| T3 | $1.1 \times 10^9$ | $9.1 \times 10^4$ |
| T5 | $2.3 \times 10^9$ | $1.1 \times 10^5$ |
| T7 | $9.6 \times 10^8$ | $1.5 \times 10^3$ |
| T2 | $1.0 \times 10^9$ | $1.7 \times 10^8$ |
| T4 | $2.2 \times 10^9$ | $1.3 \times 10^9$ |
| T6 | $1.7 \times 10^9$ | $1.1 \times 10^9$ |
| Ox1 | $6.1 \times 10^8$ | $3.0 \times 10^9$ |
| Ox3 | $1.1 \times 10^9$ | $9.2 \times 10^8$ |
| Ox5 | $1.4 \times 10^9$ | $8.0 \times 10^8$ |
| Ox6 | $8.3 \times 10^8$ | $5.2 \times 10^8$ |
| $\phi \times 174$ | $2.0 \times 10^5$ | $2.0 \times 10^5$ |
| S13 | $5.0 \times 10^6$ | $5.0 \times 10^6$ |
| MS2 | $8.0 \times 10^8$ | $1.0 \times 10^4$ |

Table 2. Inactivation of several viruses by oxidized spermine

| | Virus | Assay | Inactivation (Log 10) |
|---|---|---|---|
| Animal viruses | Sindbis | Plaque | 4.3* |
| | West Nile | Plaque | 4.4 |
| | Vaccinia | | 0.4 |
| | Newcastle disease | Plaque | 1.9 |
| | Newcastle disease | Egg | 1.7 |
| | Influenza virus | Egg | 2.7 |
| Plant viruses | TMV-Tobacco Mosaic virus | Plant | 1.07† |
| | AMV-Alfalfa Mosaic virus | Plant | 1.11 |
| | PVX-Potato virus X | Plant | 0.74 |

* Oxidized spermine, 100 µg/ml, was incubated with the respective viruses for 3 hr at 37°C and assayed.
† Oxidized spermine, 600 µg/ml, was incubated with the respective viruses for 90 min at 37°C and assayed.

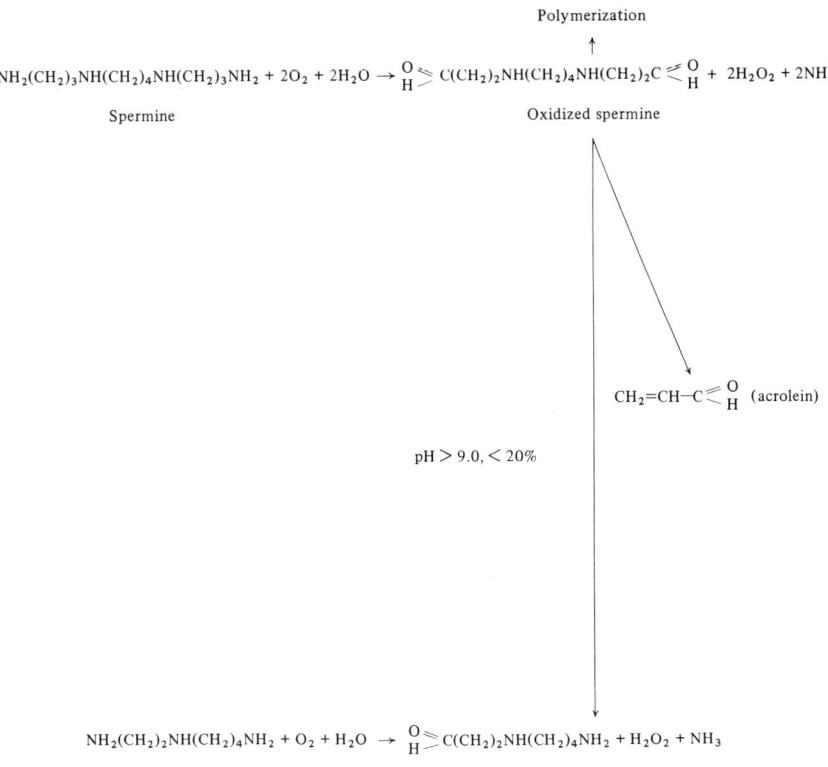

Figure 1. Oxidation of polyamines by serum amine oxidase.

## MODE OF ACTION

Subsequent studies showed that the carbonyl groups of the aminoaldehydes are required for the antiviral effect, since reduction of oxidized spermine by sodium borohydride abolished toxicity [12, 19]. More detailed studies demonstrated that oxidized spermine had a high affinity for bihelical nucleic acids and that it was bound to DNA through electrostatic and covalent bonds [20]. The reversible electrostatic binding was due to the interaction of the secondary amino groups of the aminoaldehyde with the negatively charged phosphates, while irreversible covalent bonds were formed between the carbonyl groups of oxidized spermine and amino group of purine and pyrimidine bases [21]. Quantitative studies showed that 3-15 molecules of oxidized spermine were bound per $10^3$ nucleotide units of native DNA [22] and that the binding of oxidized spermine to the various bases was in the decreasing order of reactivity: guanine, cytosine adenine and thymine. As expected, oxidized spermine induced the formation of cross-links between complementary polynucleotide chains. This was demonstrated by significant effects on thermal denaturation profiles [20] and by inhibition of DNA replication and transcription [23]. Inhibition of RNA synthesis by oxidized spermine was used by us to determine the life-time of different mRNA species [24]. Results of these experiments are summarized in Table 3.

Table 3. Physical and biological effects of oxidized spermine

A. Binding
  1. *Affinity for polynucleotide chains*
    (Two distinct bonds)
  2. *Order of reactivity*
    guanine > cytosine > adenine > thymine
  3. *Binding capacity*
    3-15 molecules per 1000 nucleotides

B. Physical effect
  1. *Stabilization of bihelical structure*
    (Denaturation profile, susceptibility to nucleases)

C. Biological effect
  1. *Inhibition of DNA transcription*
    (Determination of mRNA life-time)
  2. *Inactivated T5 phage*
    Adsorbs normally
    Injects DNA (Coupled with oxidized spermine)
    Injected DNA does not replicate and does not transcribe

After establishing the effect of oxidized spermine on nucleic acids, we attempted to clarify its antiviral action. The phage-bacteria system was chosen because of its relative simplicity and because of the detailed information available regarding the various steps of infection.

The following explanations can be offered to explain the inactivation of bacterial viruses: (a) interference of the drug with the adsorption of the inactivated virus to its host; (b) effect of the drug on the injection of viral nucleic acid into the host; and (c) modification of viral nucleic acids by the inactivating agent.

The first possibility was ruled out by various experiments including the use of fluorescent antibodies, electron microscope and labelled T5 phages. All these experiments indicated that inactivated phages adsorb to their host with an efficiency greater than 80% with respect to the untreated controls [25].

Blending experiments according to Hershey and Chase [26] showed that inactivated phages injected their DNA into the hosts. In these experiments we used $^{32}$P-labelled T5 phages, which were inactivated by $^3$H-oxidized spermine. These experiments also showed that the injected $^{32}$P-labelled DNA was coupled with $^3$H-oxidized spermine [25]. The association of oxidized spermine with the injected viral DNA was also confirmed by the extraction of the double labelled complex from the infected cells. Even though oxidized spermine had no effect on the transfer of phage DNA into the bacterial cell, viral nucleic acids lost their biological activity by interacting with oxidized spermine. Thus, only bacterial DNA and no phage DNA was formed in *E. coli* cells infected with oxidized spermine-treated T5 phages [25]. Similarly, mRNA species formed after infection with inactivated viruses were complementary to bacterial and not to phage DNA. It may be inferred that neither replication nor transcription took place with oxidized spermine-treated DNA.

## INACTIVATION OF ANIMAL VIRUSES BY OXIDIZED SPERMINE

During the recent year our attention was focused on the effect of oxidized spermine on myxoviruses and on vaccinia viruses. We demonstrated that this compound was more active against different viruses as compared to other mono and dialdehydes. Inactivation of myxoviruses was studied either by assaying the production of haemagglutinins in chick embryos or by plaque titration on chick embryo fibroblast cultures. Haemagglutinin production was assayed when viruses were first serially diluted in nutrient broth and then incubated with oxidized spermine for various times and finally propagated in chick embryos. It could be shown [19] that Sendai and influenza PR8 were more susceptible to inactivation by oxidized spermine in comparison to NDV. Fig. 2 shows that undiluted Sendai or influenza PR8, treated with oxidized spermine, 1.65 μmoles/ml for 3 to 4 hr, did not produce haemagglutinins, while NDV required 6 to 8 hr to yield similar results. Additional experiments showed (Fig. 3) that inactivation was proportional to the concentration of the drug. Again, influenza was more sensitive than NDV. Subsequent experiments showed that the inactivation of influenza virus by oxidized spermine was temperature dependent and maximal effect was obtained at 30°C (Fig. 4).

Inactivation of myxoviruses was not limited to oxidized spermine; related amino aldehydes exerted comparable effects. Thus, oxidized spermidine also inactivated these viruses but higher concentrations of the drug were required (Fig. 5). Again, NDV was more resistant to the drug than influenza PR8 when tested by haemagglutin production in embryonated eggs. Synthetic aminodialdehydes behaved like oxidized spermine, and 0.8 μmole/ml completely inactivated influenza PR8 after exposure for 24 hr at 37°C (Fig. 6).

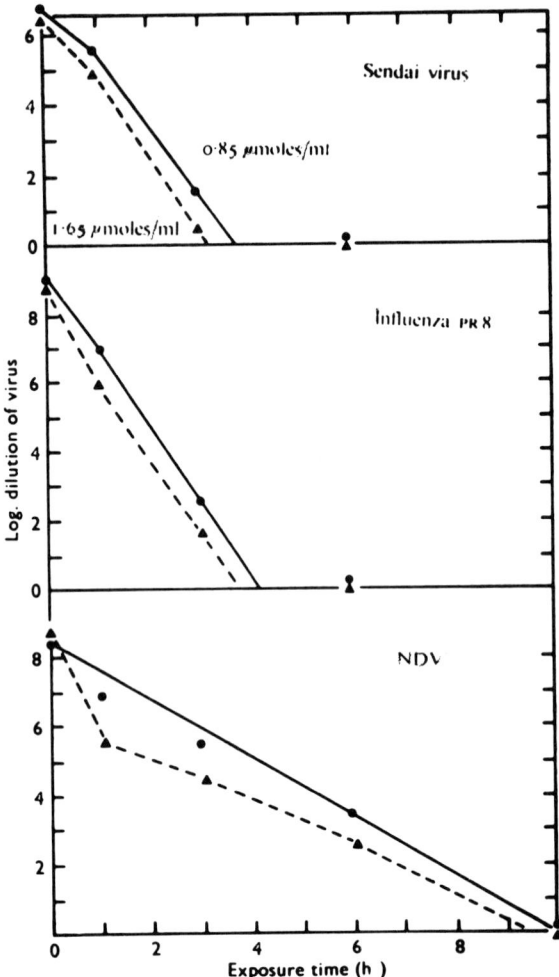

**Figure 2.** Effect of oxidized spermine on haemagglutinin production. Results are expressed as log dilution of viruses which did produce haemagglutinins after propagation in chick embryos.

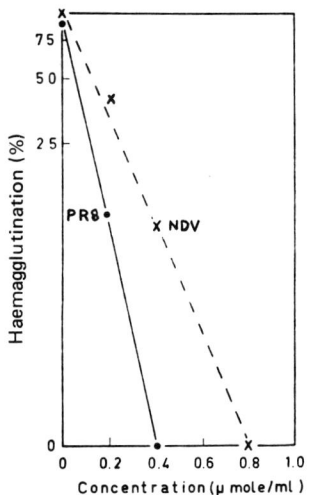

**Figure 3.** Effect of drug concentration on haemagglutinin production. Results are expressed as percentage of haemagglutination titres relative to control. Viruses were incubated with the drug for 24 hr and then propagated in chick embryos.

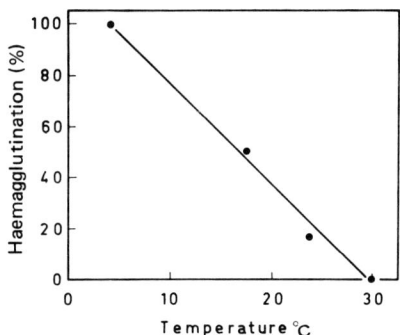

**Figure 4.** Effect of temperature on haemagglutinin production. Oxidized spermine (1.65 μmoles/ml) was incubated with influenza PR8 for 24 hr and then propagated in chick embryos.

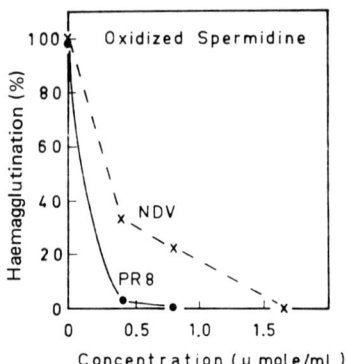

**Figure 5.** Inactivation of myxoviruses by oxidized spermidine. Experimental conditions were as given in Fig. 3, except that oxidized spermidine was used.

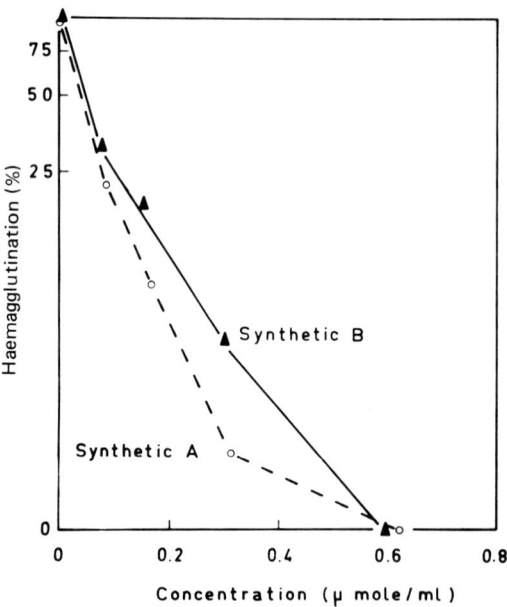

**Figure 6.** Effect of synthetic aminoaldehydes on the infectivity of influenza virus.

(A) $\overset{O}{\underset{H}{\nwarrow}}C(CH_2)_2NH(CH_2)_4NH(CH_2)_2C\overset{O}{\underset{H}{\nearrow}}$

(B) $\overset{O}{\underset{H}{\nwarrow}}CCH_2NH(CH_2)_4NHCH_2C\overset{O}{\underset{H}{\nearrow}}$

## ANTIVIRAL ACTION OF OTHER ALDEHYDES

Numerous studies have demonstrated the toxicity of aldehydes for various viruses. It was therefore of interest to find out whether the antiviral effect of oxidized spermine differed quantitatively and qualitatively from that of other aldehydes. In the following experiments NDV and vaccinia virus were incubated at 37°C with various aldehydes and their infectivities determined by the plaque assay. It may be seen (Fig. 7) that oxidized spermine and oxidized spermidine at

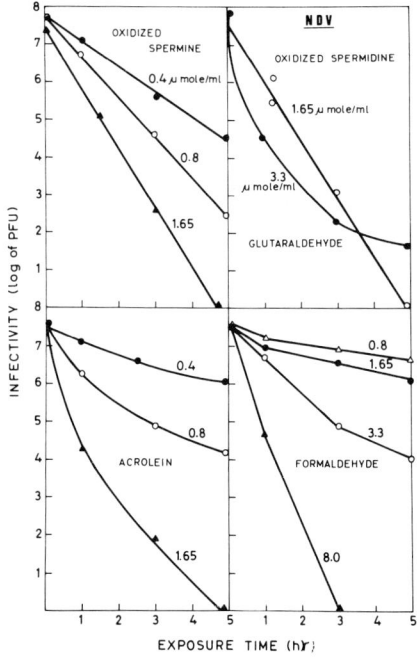

**Figure 7.** Effect of oxidized polyamines and other aldehydes on the infectivity of NDV. Viruses were incubated with the aldehydes for various times and infectivity determined by the plaque assay.

a concentration of 1.65 μmoles/ml completely inactivated a suspension of NDV within 5 hr. Glutaraldehyde and formaldehyde were less active and even at high concentration (3.3 μmoles/ml) the inactivation was not complete. The toxicity of acrolein for NDV was somewhat lower than that of oxidized spermine (Fig. 7). Similar results were also obtained with vaccinia virus (Fig. 8). Again, oxidized spermine caused a significant decrease in infectivity, while glutaraldehyde and formaldehyde were less active at comparable molar concentrations. Acrolein, on the other hand, resembled oxidized spermine in its ability to inactivate vaccinia virus (Fig. 8).

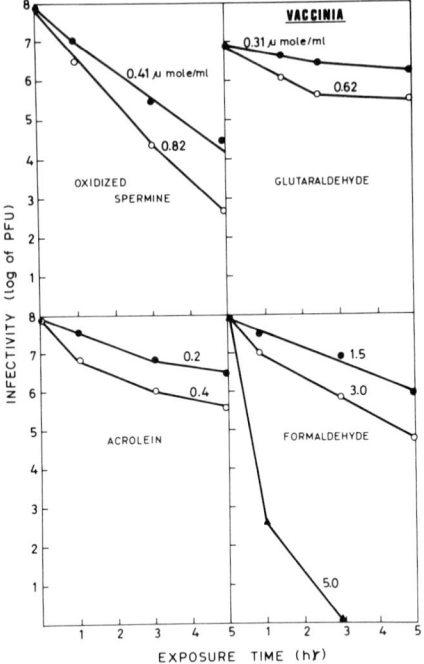

Figure 8. Effect of oxidized spermine and other aldehydes on the infectivity of vaccinia virus.

## BIOLOGICAL PROPERTIES OF INACTIVATED VIRUSES

After showing that oxidized polyamines inactivated various animal viruses, it was of interest to study the mode of its action. As mentioned above, studies with inactivated phages indicated that oxidized spermine penetrates into the susceptible phage and interacts with its nucleic acid, while only a slight amount of the drug is bound to the viral envelope. A similar binding was also shown for influenza virus: $^3$H-oxidized spermine penetrated into the virion and formed an extractable complex with its $^{14}$C-uracil-labelled RNA. If oxidized spermine inactivates viruses by interacting with their nucleic acids and not with their envelopes, inactivated virus should retain normal surface properties. The following experiments indicated that this indeed was the case.

*Haemagglutination.* Table 4 shows that myxoviruses inactivated by oxidized polyamines retained their ability to agglutinate the red blood cells. In general, inactivation of these viruses by other aldehydes impaired their haemagglutinability. It may thus be concluded that oxidized polyamines inactive myxoviruses without affecting their haemagglutinins, unlike the other aldehydes.

*Cell fusion.* Some members of the myxovirus family possess the ability to fuse certain animal cells and form giant polycaryocytes. This property serves as a sensitive parameter to study viral surface properties, since it involves enzymes and structural elements. We were able to demonstrate that Sendai virus, inactivated by oxidized spermine, retained its ability to fuse Ehrlich ascites cells [5, 19], implying the intactness of the viral envelope.

Table 4. Effect of various aldehydes on the haemagglutinability of myxoviruses *in vitro*

|  |  | Haemagglutination (%) | |
|---|---|---|---|
|  |  | Influenza PR8 | NDV (v strain) |
| Oxidized spermine | 1.65 μmoles/ml | 80 | 88 |
| Oxidized spermidine | 3.30 μmoles/ml | 100 | 75 |
| Formaldehyde | 0.05% | 21 | 44 |
|  | 0.025% | 28 | 56 |
| Glutaraldehyde | 3.30 μmoles/ml | 19 | 50 |
|  | 1.65 μmoles/ml | 50 | 100 |

*Electron microscopy.* Electron microscopy also serves as a sensitive tool to determine the intactness of a virus. These studies clearly showed that oxidized polyamines had no apparent deleterious effects on influenza PR8 and NDV particles, which appeared as intact as the untreated controls. On the other hand formaldehyde (0.05%) affected the integrity of the viruses.

## THERAPEUTIC EFFECT AND TOXICITY

Since oxidized spermine proved to be an effective antiviral agent *in vitro*, experiments were carried out to see if a parallel effect could be accomplished *in vivo* [26]. Mice, 9-11 g, were injected intraperitoneally with influenza PR8 ($5LD_{50}/0.5$ ml) and treated intranasally with crude oxidized spermine (0.05 ml, 1.32 μm/ml) twice a day, starting 36 hr prior to the infection. Thereafter, intraperitoneal treatment (0.5 ml) was continued for another nine days. It may be seen (Fig. 9) that oxidized spermine-treated mice showed a mortality of 28%, compared to 68% of the untreated infected mice, or a 40% increase in survivors versus saline-treated mice. It is also of interest that the PR8-infected, oxidized spermine-treated mice did not lose weight and the lungs of the survivors showed lesions and slight consolidations that certified that, although viral infection had taken place, it had been arrested. It can also be inferred from this experiment that oxidized spermine is not toxic for mice when given in the above described routes and quantities. This was also verified in another experiment in which oxidized spermine (0.2 ml-0.66 µmole) was injected into the allantoic cavity of embryonated eggs and examined four days later. The results (Table 5) showed no significant differences between the embryos of untreated controls and the drug-inoculated eggs.

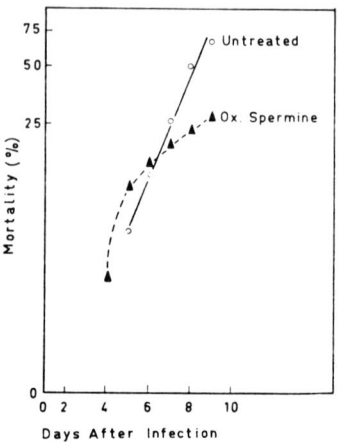

**Figure 9.** Effect of oxidized spermine on influenza-infected mice. Mice were infected intranasally and treated intraperitoneally with oxidized spermine or saline twice starting 36 hr prior to infection with PR8 ($5LD_{50}/0.05$ ml). Intraperitoneal treatment was continued after infection. The results are the cumulative of three experiments.

**Table 5.** Effect of oxidized spermine on the chick embryo

|  | Mortality (%) | Average Weight (grams) | Feathers | Deformities |
| --- | --- | --- | --- | --- |
| Oxidized spermine | 10 | 9.2 | Normal | None |
| Control | 10 | 7.0 | Normal | None |

Groups of 10 9-day old chick embryos were injected (allantoic cavity) with oxidized spermine (0.66 µmole/0.2 ml) or left untreated. Four days later the embryos were examined.

## CONCLUSION

Oxidized polyamines were shown to inactivate various bacterial, plant and animal viruses. The drug apparently penetrates into the virion and binds to its nucleic acid. This was demonstrated for both T5 phage and influenza PR8. Inactivated viruses adsorb to their host but apparently the expression of the genetic material is inhibited. Oxidized spermine appears to be more active than glutaraldehyde and formaldehyde and relatively non-toxic for animals. This last property may perhaps permit its application as a prophylactic antiviral agent.

Work is now in progress to determine the optimal structure for an active amino aldehyde, using synthetic compounds.

## ACKNOWLEDGEMENT

This study was supported by a grant from The Wellcome Trust.

## REFERENCES

1. Tabor, H. and Tabor, C. W. (1964). *A. Rev. Pharmac.* **16**, 245.
2. Bachrach, U. (1970). *A. Rev. Microbiol.* **24**, 109.
3. Liquori, A. M., Constantino, L., Crescenzi, V., Elia, V., Giglio, E., Puliti, R., De Santis Savino, M. and Vitagliano, V. (1967). *J. molec. Biol.* **24**, 113.
4. Suwalsky, M., Traub, W., Shmueli, U. and Subirana, J. A. (1969). *J. molec. Biol.* **42**, 363.
5. Bachrach, U. (1970). *Ann. N.Y. Acad. Sci.* **171**, 939.
6. Tabor, C. W., Tabor, H. and Bachrach, U. (1964). *J. biol. Chem.* **239**, 2194.
7. Kimes, B. W. and Morris, D. R. (1971). *Biochim. biophys. Acta* **228**, 223.
8. Alarcon, R. A. (1966). *Archs Biochem. Biophys.* **113**, 281.
9. Alarcon, R. A. (1970). *Archs Biochem. Biophys.* **137**, 365.
10. Bachrach, U. and Persky, S. (1964). *J. gen. Microbiol.* **37**, 195.
11. Bachrach, U., Tabor, C. W. and Tabor, H. (1963). *Biochim. biophys. Acta* **78**, 768.
12. Bachrach, U. and Leibovici, J. (1965). *Israel J. Med. Sci.* **1**, 541.
13. Fukami, H., Tomida, I., Morino, T., Yamada, H., Oki, T., Kawasaki, H. and Ogata, K. (1967). *Biochem. biophys. Res. Commun.* **28**, 19.
14. Oki, T., Kawasaki, H., Ogata, K., Yamada, H., Tomida, I., Morino, T. and Fukami, H. (1968). *Agric. biol. Chem.* **32**, 1349.
15. Oki, T., Kawasaki, H., Ogata, K., Yamada, H., Tomida, I., Morino, T., Fukami, H. (1969). *Agric. biol. Chem.* **33**, 994.
16. Bachrach, U., Rabina, S., Loebenstein, G. and Eilon, G. (1965). *Nature, Lond.* **208**, 1095.
17. Katz, E., Goldblum, T., Bachrach, U. and Goldblum, N. (1967). *Israel J. Med. Sci.* **3**, 575.
18. Bachrach, U. and Don, S. (1970). *Israel J. Med. Sci.* **6**, 435.
19. Bachrach, U. and Don, S. (1971). *J. gen. Virol.* **11**, 1.
20. Bachrach, U. and Eilon, G. (1967). *Biochim. biophys. Acta* **145**, 418.
21. Eilon, G. and Bachrach, U. (1969). *Biochim. biophys. Acta* **179**, 464.
22. Bachrach, U. and Eilon, G. (1969). *Biochim. biophys. Acta* **179**, 473.
23. Bachrach, U. and Persky, S. (1969). *Biochim. biophys. Acta* **179**, 484.
24. Bachrach, U. and Persky, S. (1966). *Biochem. biophys. Res. Commun.* **24**, 135.
25. Bachrach, U. and Leibovici, J. (1966). *J. molec. Biol* **19**, 120.
26. Hershey, A. D. and Chase, M. (1952). *J. gen. Physiol.* **36**, 39.
27. Don, S. (1970). M.Sc. thesis, Hebrew University, Jerusalem, Israel.

## DISCUSSION

**B. Rada:** What is the toxicity of oxidized spermidine on mammalian cells. Is there an appreciable difference between the concentration toxic for the host cells and that required for virus multiplication, i.e. is there some selective action?

**U. Bachrach:** The virus is affected only before it enters the cell.

**B. Rada:** Some synthetic viral inhibitors (benzimidazoles, nucleoside analogues) do not exhibit direct effects on viruses. On the other hand, substances with direct effects are not active *in vivo*. What was the virus dose in animal experiments? Was the treatment prophylactic?

**U. Bachrach:** The virus dose was 10 $LD_{50}$. Prophylactic treatment was applied for 40 hr.

**J. Doskocil:** What portion of the T5-DNA is injected? Can you correlate the number of lethal hits with the number of a particular type of lesion in the viral nucleic acid?

**U. Bachrach:** Following treatment of T5 phage with oxidized spermine, the entire phage genome is transferred. This may be explained by the formation of cross-links between the two DNA strands which are known to contain breaks. The cross-linking material may thus "repair" the gaps and fulfil the function of a ligase. As regards the second question, this has not yet been done.

**P. Langen:** The mode of action of the oxidized polyamines is similar to that of alkylating agents insofar as they cross-link the DNA strands. Is there any cross resistance with alkylating agents which would indicate that the damage inflicted by the oxidized polyamines could be repaired?

**U. Bachrach:** We tested this possibility in a bacterial system. Ultraviolet resistant mutants of *E. coli*, which are known to possess active repair mechanisms, are more resistant to oxidized spermine as compared to the wild type, pointing to the analogy of the two systems.

**E. De Clercq:** What is the direct effect of oxidized polyamines, such as oxidized spermine, on the infectivity of the viral nucleic acid itself, as compared to their effect on the intact virus particle?

**U. Bachrach:** This has been tested with phage systems. MS2 ribonucleic acid, for example, lost its infectivity after treatment with oxidized spermine.

**L. Thiry:** You showed that oxidized spermine inactivates arboviruses to a greater extent than myxoviruses and claimed that this was due to differences in permeability barriers. Could it not be due to the fact that the RNA of arboviruses, *in situ*, possess a higher degree of secondary structure?

**U. Bachrach:** Permability barriers have been demonstrated in the coliphage system. A permeable mutant of T4 (T4-0) is sensitive to oxidized spermine, in contrast to the wild type. Secondary structures may also contribute to the observed antiviral effect.

**O. P. van Diggelen:** Have oxidized spermine and spermidine been tested for ability to induce interferon?

**U. Bachrach:** They have not been tested.

**D. Shugar:** You mentioned that bound polyamines can be replaced from DNA by ions. Are oxidized polyamines also reversibly bound and replaced by ions? If so, would you not expect phage DNA (from phage previously treated with oxidized polyamines) to regain activity following entry into the cell?

**U. Bachrach:** No, because oxidized spermine is bound only irreversibly to nucleic acids, as a result of the formation of Schiff bases between the carbonyl groups of the drug and the amino groups of the bases.

# A Molecular Orbital Approach to the Study of Some Stages of Purine Metabolic Pathways

JOSÉ KANETI and EVGENY GOLOVINSKY

*Biochemical Research Laboratory, Bulgarian Academy of Sciences, Sofia, Bulgaria*

and

STEFKA DIMOVA

*Mathematical Institute with Computer Centre, Bulgarian Academy of Sciences, Sofia, Bulgaria*

The purpose of this paper is to introduce a new approach to the study of the structure-activity relationships for biologically active compounds, whose development commenced in the last ten years. This approach is inseparably linked to the development of quantum chemistry and its application to studies of molecular structure, properties and interactions, and it is hoped that it will provide more detailed information on the course and regulation of biochemical processes. In particular, this method of approach may be of some value in the design of specific enzyme inhibitors intended to serve as anti-viral and/or anti-tumour agents.

Some encouraging results have already been attained in this field with regard to both theory and practice. A number of symposia on quantum biochemistry and electronic aspects in biochemistry have been held, beginning with the Ravello symposium in September, 1963. Applications of quantum chemical methods in chemical pharmacology were discussed at the Seattle symposium (October, 1969). Three separate lectures concerning problems of quantum biochemistry and theoretical pharmacology were presented at the Menton Modern Quantum Chemistry Symposium (July, 1970).

Several different quantum chemical treatments of chemical reactivity are known. One of these, due to Klopman [1, 2, 3], is based on a polyelectronic perturbation theory. According to this treatment, the free energy change for an interaction between two reactants approaching each other in a given solvent is represented by the equation

$$\Delta G = -q_r q_s \frac{e^2}{\epsilon R} + 2 \sum_{\substack{n \\ \text{occ}}} \sum_{\substack{m \\ \text{unocc}}} \frac{(C_r^n)^2 (C_s^m)^2 \beta^2}{E_m - E_n} \quad (1)$$

where $q_r$ is the charge on atom $r$, and $q_s$ on atom $s$; $\epsilon$ is the dielectric constant; $R$ is the distance between $r$ and $s$; $C_r^n$ is the coefficient of the atomic orbital $r$ in the molecular orbital of energy $E_n$ and $\beta$ is the resonance integral of the bond forming between $r$ and $s$. Two types of reaction control can be discerned on the basis of the significance of the first or the second term of (1)—charge respectively frontier controlled [2, 4]. On the other hand, it is known that

$$\Delta G^\circ = -2{,}3\, RT \log K^\circ \qquad (2a)$$

for an equilibrium and

$$\Delta G^\ddagger = -2{,}3\, RT \log K - 2{,}3\, RT \log \frac{\kappa\, kT}{h} \qquad (2b)$$

where $\Delta G^\ddagger$ is the free energy of activation, $\kappa$ is a transmission factor, $k$ is the Boltzmann constant and $h$ is the Planck constant. Thus, one can anticipate that every series of elementary interactions should obey a correlation equation of the type

$$\log K = A \cdot f(C_r) + B \qquad (3)$$

since the charges $q_r$ are also functions of the molecular orbital coefficients ($K$ is the equilibrium or the rate constant for each compound of the considered reaction series and $f(C_r)$ is some function of the molecular orbital coefficients of atom $r$ participating in the interaction). Several different $f(C_r)$ are called reactivity indices and are frequently used in practice to relate chemical reactivity to molecular orbital characteristics.

The approach introduced above is evidently closely related to the empirical approach in the interpretation of organic chemical reactivity based on the linear free energy principle. This principle states that the values of $\Delta G$ due to structure alterations in a given reaction series are linearly related to the free energy changes in any reaction series with the same variable parameters [see e.g. refs 5, 6, 7]. Mathematically this principle is represented by an equation of the type

$$\log \frac{K_s}{K_o} = \rho\sigma \qquad (4)$$

where $K_s$ is the rate or the equilibrium constant for a substituted compound; $K_o$ is the same constant for the parent unsubstituted compound; $\rho$ is a reaction constant and $\sigma$ is a substituent constant. This equation was first introduced by Hammett [8].

The linear free energy principle is extensively used for solution of structure-chemical reactivity problems. Numerous applications of the same principle to structure-biological activity problems have been described by Hansch et al. [9].

These considerations provide a sufficient basis for application of the data obtained by molecular orbital calculations for biologically active compounds to enzymic reaction studies. The electronic reactivity indices, compared with the empirical Hammett type constants, have the obvious advantage that they represent explicitly the characteristics of the reaction centre. Thus, the use of electronic reactivity indices is particularly desirable in studies of enzyme-substrate interactions since weak noncovalent bond formation is frequently involved in the latter. Apparently the analysis of correlations between electronic characteristics and such enzymic characteristics as substrate or inhibitor constants is the only indirect method proper to study the localization of the above interactions and to obtain information concerning the active sites of enzymes being of interest.

The practical significance of such investigations is that these provide extremely valuable information concerning the behaviour of biologically active compounds and allow quantitative prediction of the biological effects of large series of chemically related compounds.

Thus, a problem of paramount importance is to be solved by quantum biochemistry—the extensive and thorough study of the enzymic stages of nucleic acid biosynthesis as well as of the metabolic processes involving separate nucleic acid components. The solution of this difficult but attractive problem can be achieved as a result of collective efforts to obtain comprehensive data for the catalysis and regulation of these processes and for quantitative treatment of these data with the aid of quantum chemical methods.

To date results from detailed studies are available for some enzymes involved in purine metabolic pathways. This allows the study of some properties of these enzymes with the aid of molecular orbital calculations. The subject to be discussed in the present paper comprises data concerning the properties of two purine phosphoribosyltransferases (adenine—EC 2.4.2.7 and hypoxanthine—EC 2.4.2.8) and of adenosine kinase (EC 2.7.1.20). One of the motives for the choice of these enzymes was that the latter are believed to be responsible for the activation of purine analogues and their nucleosides used in chemotherapy [10].

## MOLECULAR ORBITAL CALCULATIONS

The pi-electronic characteristics of the ground states of some analogues of adenine, hypoxanthine and guanine as well as of the base components of some analogues of adenosine were calculated with the aid of the semi-empirical LCAO SCF MO method of Pariser and Parr [11], and Pople [12]. Parameters proven to give satisfactory results in calculations for various purines and pyrimidines and listed by Ladik and Biczo [13] were employed.

The molecular geometry of adenine analogues was assumed to be identical with that of adenine [14]. The analogues of hypoxanthine and guanine were assumed to possess bond lengths and angles identical with these reported for guanine [15].

The numerical solutions of the self-consistent eigenvalue problems were obtained with the aid of a MINSK computer in the Mathematical Institute with Computer Centre of the Bulgarian Academy of Sciences.

## PROPERTIES OF PURINE PHOSPHORIBOSYLTRANSFERASES

The properties of two purine phosphoribosyltransferases are comparatively well known. The reactions catalysed by these enzymes are

$$\text{Adenine + PP-ribose-P} \rightleftharpoons \text{Adenylic acid + PP}_i$$

and $\quad$ Hypoxanthine + PP-ribose-P $\rightleftharpoons$ Inosinic acid + $PP_i$

Evidence has been obtained that some preparations of hypoxanthine phosphoribosyltransferase, e.g. from human erythrocytes [16], use as substrates hypoxanthine and guanine and that the binding site for both substrates is the same.

The equilibrium position of the above reactions lies far to the right [17]. The available data for the substrate specificity of purine phosphoribosyltransferases are concerned mainly with the binding of various inhibitors to the corresponding enzyme. Enzymic conversion rates have been determined only for some individual analogues.

The structure requirements for substrates or inhibitors of these two enzymes may be briefly summarized as follows. Adenine is the most tightly bound substrate of adenine phosphoribosyltransferase. Analogues of adenine are at least 40-fold more weakly bound to this enzyme. According to the data of Krinitsky et al. [18], the 6-amino group and the ring nitrogen atoms are essential for binding with the enzyme. Apparently the largest role is played by atom $N_3$ and smaller contributions to the binding come from $N_9$ and $N_7$. No data concerning the role of $N_1$ have been reported [18].

Adenine phosphoribosyltransferase is highly specific with regard to substituents in position 2 of the inhibitors. Small and hydrophobic substituents in this position are required for effective binding. It has been pointed out that the bulk of substituents attached to $C_8$ of the purine ring exerts little effect on the binding [18]. Analogues of adenine altered in the imidazole ring possess very different affinities for adenine phosphoribosyltransferase. For example, 4-aminopyrazolo-/3,4-d/-pyrimidine is the best bound analogue while 8-aza-adenine has a very weak binding affinity. 7-Deaza-adenine also binds rather weakly to the enzyme [18].

The above data were compared with the calculated electronic characteristics of the corresponding adenine analogues (Table 1). It may be seen that the calculated electronic densities of $N_7$ and $N_9$ are practically equal for the considered compounds. Apparently a $Q_7$ larger than 1.139 is proper for effective binding since the compounds with lower electronic density on $N_7$ do not bind significantly to the enzyme.

No relation can be observed between the electronic densities on the ring $N_1$ and $N_3$ atoms and the binding to adenine phosphoribosyltransferase. For 2-substituted adenines, however, the inhibitor constant $K_i$ decreases in the series 2-$CH_3$, 2-$CH_3S$, 2-F substituted compounds, i.e. parallel to the enhancement of electronic densities on $N_1$ and $N_3$. Because of the high specificity of the enzyme to the bulk of the substituent at this position, more data are not available and no quantitative relation between $Q_1$ and $Q_3$ on the one hand and $K_i$ on the other was found.

Table 1. Electronic characteristics of some adenine analogues, experimental values of $K_i$ and observed rates of the enzymic reaction catalysed by adenine phosphoribosyltransferase. $Q_r$ — the electronic density on atom $r$; $f_9^e$ — the frontier electrophilic electron density on $N_9$ (× $10^{-4}$); $S_9^e$ — the electrophilic superdelocalizability of $N_9$ (× $10^{-4}$); $K_i$ — the inhibitor constant (× $10^{-5}$ M).

| Compound | $Q_1$ | $Q_3$ | $Q_7$ | $Q_9$ | $f_9^e$ | $S_9^e$ | $K_i$ | $V_{obs}$ |
|---|---|---|---|---|---|---|---|---|
| Purine | 1.122 | 1.136 | 1.140 | 1.876 | | | 46 | |
| Adenine | 1.148 | 1.147 | 1.139 | 1.878 | 82 | 1203 | 0.069 | 50 |
| 6-Methylpurine | 1.120 | 1.134 | 1.141 | 1.876 | | | 58 | |
| 6-Carboxypurine | 1.082 | 1.110 | 1.137 | 1.876 | | | >100 | |
| 6-Chloropurine | 1.141 | 1.145 | 1.139 | 1.878 | 77 | | >100 | |
| 6-Methoxypurine | 1.147 | 1.147 | 1.138 | 1.874 | | | >100 | |
| 6-Methylthiopurine | 1.132 | 1.140 | 1.139 | 1.878 | 69 | | >100 | |
| 2,6-Diaminopurine | 1.169 | 1.174 | 1.138 | 1.878 | 72 | | >100 | 0.43 |
| Hypoxanthine | 1.883 | 1.121 | 1.106 | 1.864 | 2.4 | 1166 | >100 | 0.22 |
| 2-Methyladenine | 1.147 | 1.147 | 1.141 | 1.876 | | | 82 | |
| 2-Chloroadenine | 1.160 | 1.162 | 1.140 | 1.876 | | | >100 | |
| 2-Fluoroadenine | 1.171 | 1.175 | 1.140 | 1.877 | 75 | | 2.8 | |
| 2-Methylthioadenine | 1.159 | 1.161 | 1.141 | 1.876 | | | 26 | |
| 8-Methyladenine | 1.148 | 1.136 | 1.135 | 1.877 | | | 5.3 | |
| 4-Aminopyrazolo/3,4-d/-pyrimidine | 1.151 | 1.147 | 0.966 | 1.876 | 197 | 1228 | 2.7 | 0.57 |
| 7-Deaza-adenine | 1.149 | 1.147 | 1.045 | 1.882 | 168 | | 49 | |
| 8-Aza-adenine | 1.125 | 1.146 | 1.061 | 1.869 | 194 | | >100 | |
| 5-Aminoimidazole-4-carboxamide | | | | | 5 | 1155 | 58 | 0.83 |

The data concerning the binding of adenine analogues altered in the imidazole ring are quite interesting. It should be mentioned that a basic $N_7$ atom is absent in these compounds. The calculated electronic densities for these compounds at atoms corresponding to $N_3$ in a purine ring and to the amino group in adenine are equal. The electronic density at the atom corresponding to $N_1$ in a purine increases in the series 8-aza-adenine, 7-deaza-adenine, 4-aminopyrazolo-/3,4-d/-pyrimidine. The binding affinity of these compounds also increases in the same sequence. This fact, as well as the behaviour of 2-substituted adenine derivates, may be considered as an indication that $N_1$ in adenine analogues also plays a significant role in the interaction of adenine phosphoribosyltransferase with its substrates and inhibitors.

The best bound substrate for hypoxanthine-guanine phosphoribosyltransferase is guanine. The substrate constants for this enzyme, however, are of the same order of magnitude for hypoxanthine, 6-mercaptopurine and 6-thioguanine [19]. It has been stated that basic (pyridine type) $N_3$ and $N_7$ atoms, as well as an acidic (pyrrole type) $N_1$ atom, are required [20] for effective binding with this enzyme. $N_9$ also plays some role in the interaction with hypoxanthine phosphoribosyltransferase. Guanine and hypoxanthine analogues altered in the imidazole ring bind comparatively weakly to this enzyme. It should be mentioned that 7-oxopyrazolo-/4,3-d/-pyrimidine, which is inhibitor, but not substrate, of hypoxanthine phosphoribosyltransferase is the best binding compound among the latter. This fact indicates that the nitrogen atom corresponding to $N_7$ in a purine may also be a pyrrole type one. This is a valuable indication for the character of the interaction between the enzyme and this fragment of the substrate or inhibitor molecule.

A qualitative conformity between the experimental values of $K_i$ and calculated electronic densities for $N_3$ and $N_7$ (Table 2) of the corresponding substrates for hypoxanthine phosphoribosyltransferase may be observed. A

Table 2. Electronic characteristics of some hypoxanthine and guanine analogues, experimental values of $K_i$ (x $10^{-6}$ M) and relative rates of the enzymic reaction catalysed by hypoxanthine-guanine phosphoribosyltransferase. $Q_r$—the electronic density on atom $r$; $S_9{}^e$—the superdelocalizability of $N_9$ atom.

| Compound | $Q_{10}$ | $Q_1$ | $Q_3$ | $Q_7$ | $Q_9$ | $S_9{}^e$ | | $V_{obs}$ * | $K$ |
|---|---|---|---|---|---|---|---|---|---|
| Hypoxanthine | 1.4018 | 1.8835 | 1.1212 | 1.1064 | 1.8643 | 0.1166 | 55 | 1.11 | |
| Guanine | 1.4056 | 1.8867 | 1.1861 | 1.1269 | 1.8647 | 0.1173 | 66 | 1.24 | |
| Xanthine | 1.3716 | 1.8990 | 1.9027 | 1.1265 | 1.8584 | 0.1132 | 0.033 | – | 2 |
| Thioguanine | 1.3370 | 1.8806 | 1.1806 | 1.1229 | 1.8631 | 0.1171 | – | 1.12 | |
| 6-Mercaptopurine | 1.3269 | 1.8768 | 1.1381 | 1.1240 | 1.8616 | 0.1162 | 49 | 0.54 | |
| 6-Amino-4-oxo-/3,4-d/-pyrazolopyrimidine | 1.4063 | 1.8863 | 1.1871 | 0.9475 | 1.8616 | 0.1193 | – | – | 3 |
| 4-Oxo-/3,4-d/-pyrazolopyrimidine | 1.4042 | 1.8821 | 1.1442 | 0.9470 | 1.8575 | 0.1186 | 1.5 | – | 9 |
| 4-Thio-/3,4-d/-pyrazolopyrimidine | 1.3303 | 1.8765 | 1.1389 | 0.9516 | 1.8575 | 0.1180 | – | – | 5 |
| 7-Oxo-/4,3-d/-pyrazolopyrimidine | 1.4037 | 1.8835 | 1.1324 | 1.8541 | 0.9665 | 0.0807 | – | – | |
| 8-Aza-guanine | 1.4007 | 1.8865 | 1.1879 | 1.0588 | 1.8514 | 0.1170 | – | 0.30 | 2 |
| 8-Aza-hypoxanthine | 1.3979 | 1.8824 | 1.1446 | 1.0528 | 1.8525 | 0.1162 | – | – | 10 |

All condensed pyrimidines are numbered as if purines
* Relative rates first column from ref. [19]; second column from ref. [24].

qualitative correspondence may be observed also between the $K_i$ values and the electronic density calculated for the atom attached to $C_6$ (separate relations for O and S, Table 2). A quantitative linear relation can be observed between $K_i$ values and electronic densities at $N_1$ separately for purines and condensed pyrimidines (Fig. 1). One can expect that similar relations could be established between other electronic indices and $K_i$ provided data were available for more substrates or inhibitors of hypoxanthine phosphoribosyltransferase.

A comparison between the properties of the two considered purine phosphoribosyltransferases places in evidence the following features. Both

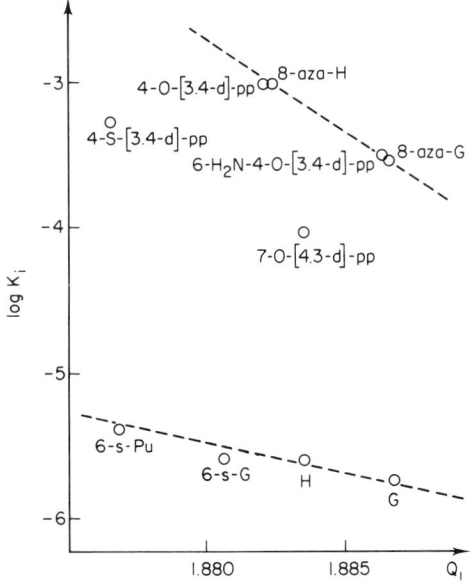

**Figure 1.** Rates of the reaction catalysed by adenine phosphoribosyltransferase against the superdelocalizability $S_9^e$ of corresponding adenine analogues.

enzymes require intact purine rings for their substrates. The values of $K_i$ for compounds with an altered imidazole ring are of the same order of magnitude for the two enzymes (about $10^{-3} - 10^{-4}$ M), those for the hypoxanthine enzyme being somewhat higher. The papers of Miller et al. [20] and Krenitsky et al. [19], however, lead to the impression that the changes in the imidazole ring of analogues cause larger effect in the binding with the hypoxanthine enzyme ($K_i$ for adenine and adenine phosphoribosyltransferase is $0.69 \cdot 10^{-6}$ M[18], while the same value for guanine and hypoxanthine phosphoribosyltransferase is $1.8 \cdot 10^{-6}$ M[19]). Another impression produced by a comparison of these two enzymes is that the hypoxanthine enzyme is more tolerant to substitution at position 2 of the purine ring. Curiously, more substituted adenines have been examined as substrates or inhibitors of adenine phosphoribosyltransferase and more condensed pyrimidines as substrates or inhibitors of the hypoxanthine enzyme.

## SOME PROPERTIES OF ADENOSINE KINASE

It has been stated that the "normal" substrate adenosine has the highest affinity with respect to adenosine kinase [21, 22]. Alterations in the base moieties of the examined nucleosides have increased the values of $K_m$ up to 50-fold. Alterations in the sugar moiety cause a considerably stronger decrease of binding with the enzyme. Essential for the binding with adenosine kinase are $N_1$ and $N_3$ of the substrates [23] and also the $2'$-OH group and apparently the $3'$-OH-group. Substituents attached to $C_{1'}$ cause loss of substrate properties [21].

Small and hydrophilic substituents attached to $C_2$ of the purine ring exert minor effects on the binding. It is essential, however, that the nitrogen atoms $N_1$ and $N_3$ be basic (pyridine type). Simultaneous alterations in both base and sugar components usually cause loss of substrate properties [21].

No simple relation between he electronic characteristics of the base moieties and experimental values of $K_m$ was found. This fact suggests that $K_m$ and $K_s$ are not identical for adenosine kinase since it would have been anticipated that $K_s$ should be simply related to the electronic characteristics of the substrates.

## ELECTRONIC STRUCTURES OF ANALOGUES AND ENZYMIC CONVERSION RATES

The effects exerted on the electronic structure of substrates by the formation of enzyme-substrate complexes can be considered in the majority of cases as small perturbations. Hence, one can use the data concerning the electronic structure of the free substrates for the interpretation of the conversion rates of enzyme-substrate complexes. The use of the Michaelis maximum velocity $V_m$ for correlations with molecular orbital indices can be justified. Frequently, however, one has to be satisfied by the presence of considerably less informative data.

The rates of purine phosphoribosyltransferase reactions have been estimated only for a small number of analogues, kinetic data for the reaction catalysed by adenine phosphoribosyltransferase being available for only five compounds including the two "normal" substrates—adenine and 5-aminoimidazole-4-carboxamide [18]. The rates of reactions catalysed by hypoxanthine phosphoribosyltransferases from different sources have been measured for 7 analogues [19, 24].

No correlation can be observed between the calculated electronic densities for $N_9$ and the velocities of phosphoribosyltransferase reactions for the corresponding compounds. Applying the equation (1) introduced by Klopman, one can see that the reaction of purines with PP-ribose-P should be a frontier but not charge controlled one. Indeed, a plot of observed rates of adenine phosphoribosyltransferase catalysed reaction against calculated superdelocalizabilities [25] for $N_9$ (Fig. 2) shows that these two quantities tend to be linearly correlated. The "normal" substrates adenine and 5-amino-imidazole-4-carboxamide, however, do not obey the tentative correlation for analogues. "Normal" substrates are converted to nucleotides significantly more rapidly than analogues. This fact may be related to the extreme affinity of adenine to the enzyme respectively to the fact that 5-amino-imidazole-4-carboxamide is not a purine. To date, however, no reasonable explanation of both the above facts can be suggested.

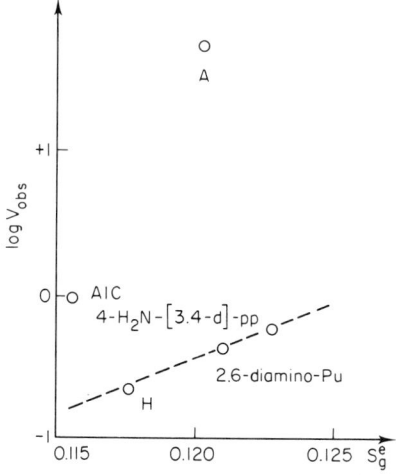

**Figure 2.** Rates of the reaction catalysed by hypoxanthine phosphoribosyltransferase against the superdelocalizabilities $S_9^e$ of corresponding hypoxanthine and guanine analogues.

A similar linear correlation between superdelocalizabilities calculated for $N_9$ and relative enzymic conversion rates exists for hypoxanthine and guanine analogues (Fig. 3). Interestingly, in this case "normal" and "abnormal" substrates form a general correlation between relative rates (log $V_{obs}$) and molecular orbital indices. It should be noted that the affinities of all good substrates to hypoxanthine phosphoribosyltransferase are of the same order of magnitude [19, 20] in contrast to the situation reported for adenine phosphoribosyltransferase [18].

One can see from Fig. 3 that the value of log $V_{obs}$ for 4-oxo-pyrazolo-/3,4-d/-pyrimidine is significantly below the correlation line established for purine analogues. The same deviation can be observed for 8-azaguanine in the reaction catalysed by enzyme obtained [24] from an 8-azaguanine resistant mutant of *Salmonella*. One can speculate that the mechanism of catalysis for purines and condensed pyrimidines is different because of the absence of the basic $N_7$ atom in the latter. This suggestion is indirectly supported by the observation that $N_7$ plays an essential role in the interaction of substrates with hypoxanthine phosphoribosyltransferase [19, 20].

The conversion rates of substrates for adenosine kinase and their relation to the electronic structure (Table 3) is somewhat more complex. Restricting consideration only to ribosides, one can see that the measured phosphorylation rates vary over a range from 0.04 to four-fold the rate for adenosine depending on the structure of the base moiety [21, 22, 23, 26]. One has no reasons,

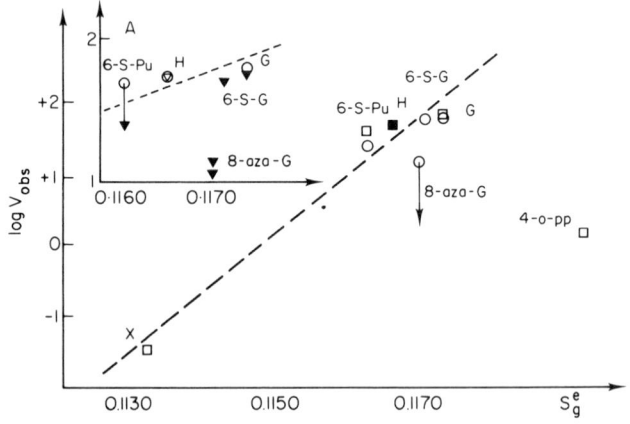

Figure 3. Inhibitor constants for some inhibitors of hypoxanthine phosphoribosyltransferase against the electronic density $Q_1$ of corresponding hypoxanthine and guanine analogues.

however, to believe that the influence of the base moiety can be transmitted to the reaction centre 5'-OH through the sugar system of sigma bonds. A more direct linkage between $N_3$ of the base and 5'-OH has to be assumed, since $N_3$ has been reported to be important for displaying substrate properties [23]. Two cases are possible—the first one is the formation of a hydrogen bond between $N_3$ and 5'-OH and base catalysis of the attack by 5'-OH on the P–O bond in ATP. One should anticipate that the reaction rate in this case will increase parallel to the basicity of $N_3$, i.e. the electronic density of $N_3$. The second possibility is the formation of a chelate complex of the cofactor $Mg^{2+}$ and $N_3$ and 5'-OH as ligands. In this case the more stable the complex, the less the reactivity of 5'-OH group in the corresponding nucleoside. Log $V_{obs}$ should decrease parallel to the enhancement of the basicity of $N_3$. The real picture of the dependence of log $V$ on the electronic density $Q_3$ of substrates is plotted in Fig. 4. The negative slope

## STUDY OF SOME STAGES OF PURINE METABOLIC PATHWAYS 173

**Table 3.** Pi-electronic characteristics of the base components of some substrates for adenosine kinase, experimental values of $K_m$ [refs 21, 22] and relative phosphorylation rates [first column—refs 21, 22; second column—ref. 23; third column—ref. 26]. All presented substrates are beta-ribosides.

| No. | Base component | $K_m \cdot 10^{-6}$ M | Relative rate | | | $Q_1$ | $Q_3$ | $f_3^e$ |
|---|---|---|---|---|---|---|---|---|
| 1 | 9H-Adenine | 1.6 | 1.00 | 1.00 | 1.00 | 1.148 | 1.147 | 0.0905 |
| 2 | Purine | 78 | 4.13 | 3.45 | 2.09 | 1.122 | 1.136 | 0.0586 |
| 3 | 6-Chloropurine | 55 | 1.02 | 2.86 | 1.77 | 1.141 | 1.145 | 0.0825 |
| 4 | 6-Methylpurine | – | – | 3.16 | – | 1.120 | 1.134 | 0.0640 |
| 5 | 6-Methylpurine | – | – | 2.09 | – | 1.147 | 1.146 | 0.0842 |
| 6 | Adenine-$N_1$-oxide | 41 | 0.62 | 4.03 | 1.08 | 1.654 | 1.137 | 0.0080 |
| 7 | 2-Fluoro-adenine | 43 | 0.21 | 2.36 | – | 1.167 | 1.171 | 0.0990 |
| 8 | 2,6-Diaminopurine | 42 | 0.04 | 0.37 | – | 1.169 | 1.174 | 0.1028 |
| 9 | 6-Methylthiopurine | – | – | 3.29 | 2.44 | 1.132 | 1.140 | 0.0709 |
| 10 | 3H-Adenine* (Isoadenosine) | 8600 | 0.80 | 0.32 | – | | 1.215 | 0.1014 |
| 11 | 8-Aza-adenine | – | – | 1.89 | – | 1.125 | 1.146 | 0.1045 |
| 12 | 7-Aminopyrazolo-/4,3-d/-pyrimidine | 220 | 0.34 | 1.01 | – | 1.137 | 1.123 | 0.2635 |
| 13 | 4-Aminopyrazolo-/3,4-d/-pyrimidine | – | – | 3.11 | – | 1.151 | 1.147 | 0.1075 |
| 14 | 3-Cyano-4-aminopyrrolo-/2,3-d/-pyrimidine | 8.0 | 1.05 | – | – | 1.148 | 1.147 | 0.0679 |
| 15 | 4-Aminopyrrolo-/2,3-d/-pyrimidine | 40 | 2.70 | 4.35 | – | 1.149 | 1.147 | 0.0760 |

* For $Q_3$ stands the value of $Q_9$.
All condensed pyrimidines are numbered as if purines.

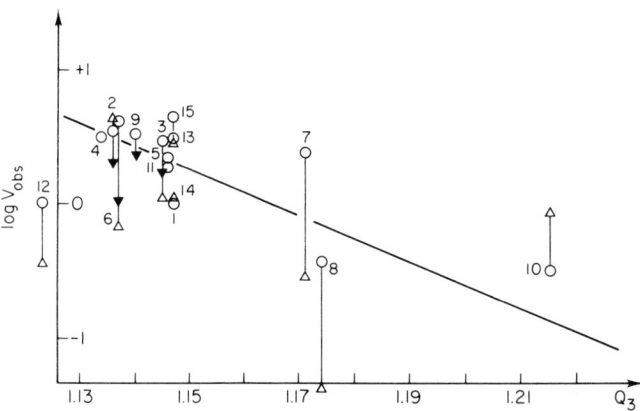

**Figure 4.** Rates of the reaction catalysed by adenosine kinase against the electronic densities $Q_3$ of corresponding base moieties. For the numbering see Table 3.

of the relation $Q_3 - \log V_{obs}$ is consistent with the suggestion that the reactive form of adenosine analogues in the reaction catalysed by adenosine kinase is the chelate with the cofactor $Mg^{2+}$. A similar correlation was established for the values of $\log V_{obs}$ and $f_3^e$ (the frontier electron density of $N_3$ [27]). A statistical treatment of the latter relation yielded the correlation equation (Fig. 5).

$$\log V_{obs} = -23.9 f_3^e + 2.1$$

which allows the satisfactory prediction of phosphorylation rates for adenosine analogues.

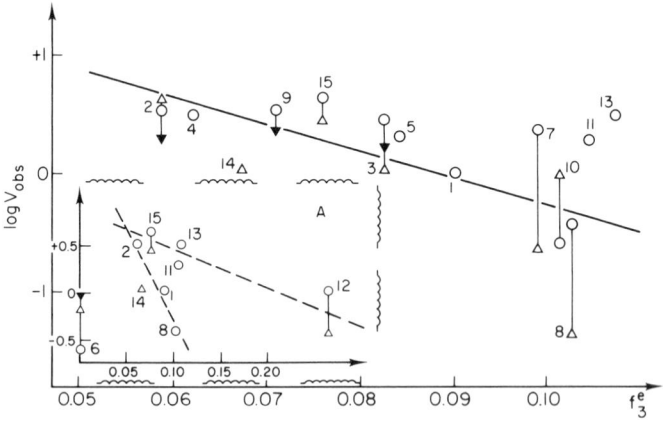

**Figure 5.** Rates of the reaction catalysed by adenosine kinase against the frontier electron densities $f_3^e$ of corresponding base moieties. For the numbering see Table 3. (A) Phosphorylation rates for beta ribosides of some condensed pyrimidines—analogues of adenine (see Table 3).

## CONCLUSION

*A priori* key metabolic reactions critical for development may be found for any type of cells. Hence, controlling a given key metabolic process with the aid of some inhibitor, one can induce proportional effects on the development of any desired organism. The only approach providing possibilities for design of inhibitors predicted to be most effective is the quantum chemical approach to the study of metabolic processes. Some experimental results verifying this statement have been reported at the Seattle Symposium on molecular orbital studies in pharmacology [28] and in a recent paper by us [29].

## ACKNOWLEDGEMENTS

We are greatly indebted to Drs I. Pojarliev, I. Juchnovski and D. Genchev for valuable discussions and advice and to Professors R. Tsanev and B. Sendov for support.

## REFERENCES

1. Klopman, G. and Hudson, R. F. (1965). *Theoret. Chem. Acta* **8**, 165.
2. Klopman, G. (1968). *J. Am. Chem. Soc.* **90**, 223.
3. Klopman, G., Tsuda, K., Louis, J. B. and Davis, R. E. (1970). *Tetrahedron* **26**, 4549.
4. Chalvet, O., Daudel, R. and McKillop, T. F. W. (1970). *Tetrahedron* **26**, 349.
5. Jaffe, H. H. (1953). *Chem. Rev.* **53**, 191.
6. Wells, P. R. (1953). *Chem. Rev.* **53**, 171.
7. Leffler, J. E. and Grunwald, E. (1963). "Rates and Equilibria of Organic Reactions," Wiley, New York.
8. Hammett, L. P. (1935). *Chem. Rev.* **17**, 125.
9. Fujita, T., Iwasa, J. and Hansch, C. (1964). *J. Am. Chem. Soc.* **86**, 5175.
10. Elion, G. B. (1969). *Cancer Res.* **29**, 2448.
11. Pariser, R. and Parr, R. G. (1953). *J. Chem. Phys.* **21**, 466, 767.
12. Pople, J. A. (1953). *Trans. Far. Soc.* **49**, 1375.
13. Ladik, J. and Biczo, G. (1969). *Acta Chim. Hung.* **62**, 401.
14. Kraut, J. and Jensen, L. H. (1963). *Acta Cryst.* **16**, 79.
15. Murayama, W., Nagashima, N. and Shimizu, Y. (1969). *Acta Cryst.* **B 25**, 2236.
16. Henderson, J. F., Brox, L. W., Kelley, W. N., Rosenbloom, F. M. and Seeiller, J. E. (1968). *J. biol. Chem.* **243**, 2514.
17. Imsande, J. and Handler, P. (1961). In "The Enzymes" (P. D. Boyer, H. Lardy and Myrback, K., eds), Vol. 5, 2nd ed., pp. 294, 296, Academic Press, New York and London.
18. Krenitsky, T. A., Neil, S. M., Elion, G. B. and Hitchings, G. H. (1969). *J. biol. Chem.* **244**, 4779.
19. Krenitsky, T. A., Papaioannou, R. and Elion, G. B. (1969). *J. biol. Chem.* **1263**.
20. Miller, R. L. and Bieber, A. L. (1969). *Biochemistry* **8**, 603.
21. Lindberg, B., Klenow, H. and Hansen, K. (1967). *J. biol. Chem.* **242**, 350.
22. Lindberg, B. (1969). *Biochim. biophys. Acta* **185**, 245.
23. Schnebli, H. P., Hill, D. L. and Benneth, L. L. (1967). *J. biol. Chem.* **242**, 1997.
24. Adye, J. C. and Gots, J. S. (1966). *Biochim. biophys. Acta* **118**, 344.
25. Fukui, K., Yonezawa, T. and Nagata, C. (1957). *J. chem. Phys.* **27**, 1247.
26. Leibach, T., Spiess, G., Neudeccker, T., Peschke, G., Puchwein, G. and Hartmann, G. (1971). *Hoppe Seyler's Z. Physiol. Chem.* **352**, 328.
27. Fukui, K., Yonezawa, T. and Shingu, H. (1952). *J. chem. Phys.* **20**, 722. Fukui, K., Yonezawa, T. and Shingu, H. (1954). *J. phys. chem.* **22**, 1433.
28. Kier, L. B., (Ed.) (1970). "Molecular Orbital Studies in Chemical Pharmacology," Springer Verlag, Berlin–Heidelberg–New York.
29. Kaneti, J. and Golovinsky, E. (1971). *Chem. Biol. Interactions* **3**, 421.

## DISCUSSION

**G. P. Galabov**: Have you any experimental data relating to the antiviral activity of specific antimetabolites, and which support your theoretical considerations?

**J. Kaneti**: Data concerning the activity of certain antimetabolites on microorganisms, experimental tumours, and some enzymic systems have been obtained in our laboratory. In my opinion these data can be readily correlated with the antimetabolic activity on host cells.

**D. Shugar**: Have you taken account, in your calculations, of the effect of some substituents on the nucleoside conformation? For example, it is known that 8-bromoguanosine is in the *syn*, and not *anti*, conformation.

**J. Kaneti**: We have not taken account of the conformation of such compounds, since they are neither substrates, nor inhibitors, of adenosine kinase. Nucleoside conformation can, however, be taken into account with the use of more advanced calculations.

# Mode of Action of Rifamycin and Aminopiperazine Derivatives on Animal Viruses and Cells

LISE THIRY and GIANCARLO LANCINI

*Institut Pasteur du Brabant, Bruxelles, Belgium*
*and*
*Gruppo Lepetit, Milano, Italy*

Semisynthetic derivatives of the antibiotic rifamycin have been prepared by attaching various side chains to the 3-C position of the naphthohydroquinone moiety of the molecule. Fig. 1 shows the chemical formulas of rifamycin derivatives which have been assayed by various authors for their action on viruses or on animal cells. Substitutions on the 3 position include formyl dimethyl hydrazone, oxime or aminopiperazine groups. The action of the latter groups alone, not condensed with rifamycin, has also been studied.

On the other hand, some authors have compared the effects of rifamycin derivatives with those of tolypomycin R and streptovaricin, which are naturally occurring antibiotics with structural similarities to rifamycins. Antibiotics with such structures are called ansamycins.

The experiments which we are going to summarize were prompted by the original observations that rifampicin inhibits the growth of vaccinia virus [1, 2], while rifamycin does not have this effect, although both drugs inhibit the growth of bacteria by binding to the DNA-dependent RNA polymerase.

## ACTION OF RIFAMYCIN DERIVATIVES ON VACCINIA VIRUS GROWTH

It now seems very probable that inhibition of vaccinia virus growth by rifampicin is mainly related to the block of envelope formation of the virion [3, 4, 5] due to lack of assembly of preformed structural polypeptides of the virion [6, 7, 8], this event being in turn probably caused by the absence of cleavage of a higher molecular weight polypeptide precursor [9, 10]. On the other hand, rifampicin shows some action on the transcription of the viral genome but the consequences on the replication of the virus are unclear; in infected cells the drug inhibits the appearance of functional virus-specific DNA-dependent RNA

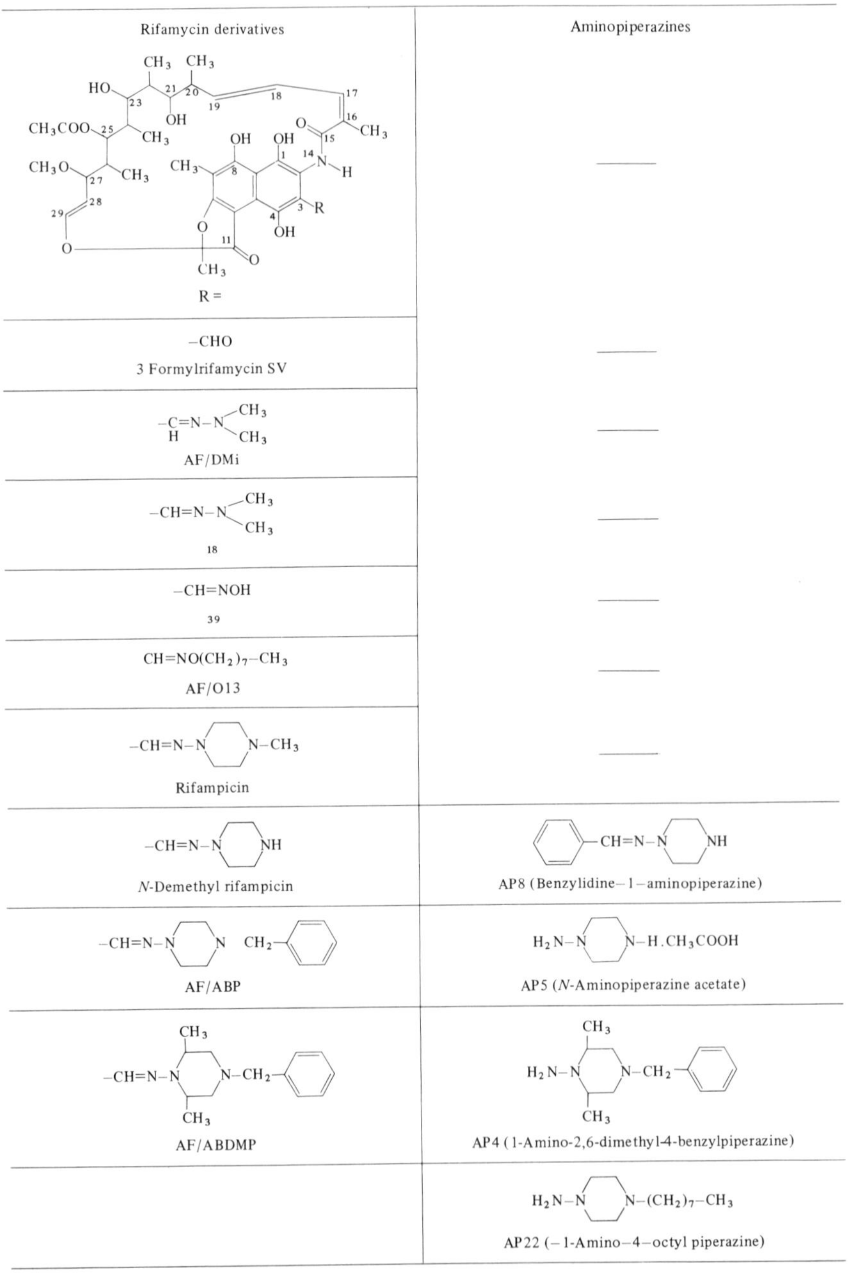

| Rifamycin derivatives | Aminopiperazines |
|---|---|
| (core rifamycin structure with positions labeled: CH₃ at 21, 20; HO at 23; CH₃COO at 25; CH₃O at 27; positions 17, 18, 19, 16, 14, 15, 11, 8, 4, 3, 1, 28, 29; OH groups; R = substituent) | |
| $-CHO$ — 3 Formylrifamycin SV | — |
| $-\underset{H}{C}=N-N\underset{CH_3}{\overset{CH_3}{\diagup}}$ AF/DMi | — |
| $-CH=N-N\underset{CH_3}{\overset{CH_3}{\diagup}}$ 18 | — |
| $-CH=NOH$ 39 | — |
| $CH=NO(CH_2)_7-CH_3$ AF/O13 | — |
| $-CH=N-N\!\!\bigcirc\!\!N-CH_3$ Rifampicin | — |
| $-CH=N-N\!\!\bigcirc\!\!NH$ N-Demethyl rifampicin | $\bigcirc\!\!-CH=N-N\!\!\bigcirc\!\!NH$ AP8 (Benzylidine–1–aminopiperazine) |
| $-CH=N-N\!\!\bigcirc\!\!N-CH_2-\bigcirc$ AF/ABP | $H_2N-N\!\!\bigcirc\!\!N-H \cdot CH_3COOH$ AP5 (N-Aminopiperazine acetate) |
| $-CH=N-N\!\!\bigcirc\!\!N-CH_2-\bigcirc$ (with CH₃ groups at 2,6) AF/ABDMP | $H_2N-N\!\!\bigcirc\!\!N-CH_2-\bigcirc$ (with CH₃ groups at 2,6) AP4 (1-Amino-2,6-dimethyl-4-benzylpiperazine) |
| | $H_2N-N\!\!\bigcirc\!\!N-(CH_2)_7-CH_3$ AP22 (−1-Amino−4−octyl piperazine) |

**Figure 1.** Formulae of rifamycin derivatives and aminopiperazines which are cited in the text.

polymerase (transcriptase) which is normally synthesized in the cytoplasm late in the infectious cycle [11], but the synthesis of virus specific mRNA is not modified during the first cycle of infection. Also rifampicin was active in an *in vitro* assay of the virus transcriptase [12]. These effects are probably not sufficient to block vaccinia virus replication, since compound 39 is a more effective inhibitor of *in vitro* transcriptase activity of the virions than rifampicin; while only rifampicin, and not compound 39, inhibits virus plaque formation at 100 $\mu$g/ml [13,14]. Furthermore, resistant mutants of vaccinia virus, which can grow in the presence of rifampicin, still possess a virion-associated RNA polymerase which is susceptible to high concentrations of rifampicin, as is the wild type [15], and the polymerases from wild type and from rifampicin-resistant mutants are inhibited to the same extent by the very active inhibitors AF/ABDP and 3-formylrifamycin [14].

Inhibition of vaccinia virus growth by some ansamycin antibiotics may be due to toxic effects on the cells, rather than to a specific action on the morphogenesis of the virus [13, 16]. The most striking example is that of AF/ABDP, which displays a toxic effect at 50 $\mu$g/ml while it does not cause a block of virus maturation at 1000 $\mu$g/ml (Table 1); a similar, although somewhat smaller discrepancy, exists with 3-formylrifamycin SV and tolypomycin. With the non-toxic drugs, virus growth inhibition is better correlated with the effect on virus maturation than with the action on the virion transcriptase. Moreover, a vaccinia virus mutant selected only for resistance to rifampicin was also resistant to the effect on morphogenesis produced by the other antibiotics [16].

Even in bacterial systems, it now appears that inhibition of the growth of RNA bacteriophages by rifampicin is due not only to some action on the host transcriptase but involves also a later effect which blocks the release of phage particles; in that case, however, virus maturation is not blocked, since infectious particles accumulate inside the bacterial cell [17].

## *IN VITRO* ACTION OF RIFAMYCIN DERIVATIVES ON POLYMERASES

Rifamycin SV and many other rifamycin derivatives inhibit bacterial transcriptases but not the vaccinia virus enzyme. Conversely, this enzyme is strongly inhibited by one rifamycin derivative (23-dehydroxy-27-demethoxy-23-27-epoxyrifamycin SV) which does not react with the bacterial polymerase [14]. Other derivatives display other specificities. AF/ABDMP is not only active on the two previously cited polymerases, but also on the reverse transcriptase (RNA-dependent DNA polymerase) of the virions of several RNA tumour viruses of murine, feline and avian origin [18]. *N*-Demethylrifampicin, an inhibitor of bacterial polymerase, also inhibits the reverse transcriptases by 50% and 80% at 100 and 250 $\mu$g/ml respectively, while concentrations of 200 $\mu$g/ml of this antibiotic have no action on the transcriptase of vaccinia virus particles. A similar concentration of the same antibiotic also decreased by 50% the activity of an RNA-dependent DNA polymerase extracted from lymphoblasts of patients with acute lymphoblastic leukaemia (ALL); inhibition was 100% with 460 µg/ml of N-demethylrifampicin. The enzyme was partially purified and its activity was RNA-dependent [19]. Drugs more active against tumour viruses reverse transcriptase were obtained by substitution of oxime groups at the 3 position (for instance AF/013); but these drugs, in contrast to those previously cited, are also inhibitors of the DNA-dependent DNA polymerases of the animal cells [20].

Streptovaricins, which also have structures similar to rifamycin, seem more active than N-demethylrifampicin on the reverse transcriptase of murine leukaemia viruses (Moloney strain), since a streptovaricin complex, composed of seven macrolides, produces a 75% inhibition of the enzyme activity at a concentration of 40 µg/ml, although complete inhibition is not obtained with higher doses [21]. There are preliminary indications that only some of the macrolides of the complex are active, and that pure products of higher activity might be obtained. It is to be noted that streptovaricins inhibit DNA-dependent RNA polymerases of bacteria, but not those of animal cells.

Molecules with structures similar to that of rifamycin thus appear to have in common the property to bind polymerases, but small variations in the structures, either in the rifamycin moiety or in the side chains, modify the specificities for the various polymerases.

## ACTION OF RIFAMYCIN DERIVATIVES ON ANIMAL CELLS

It has already been mentioned that some drugs, such as 3-formylrifamycin and AF/DMI, exhibit cytotoxic effects: these are accompanied by mitochondrial swelling with distortion of cristae [16]; it is not known whether the lesions are related to some action on mitochondrial transcriptases, which are susceptible *in vitro* to rifampicin and to other derivatives active on bacterial transcriptase. It may be that only some of these derivatives, like AF/DMI and 3-formylrifamycin, are active on mitochondria *in vivo* because only these penetrate into the mitochondrial particles.

Apart from these toxic effects, which have not been shown to affect some types of cells in particular, N-demethylrifampicin was found to inhibit the growth of hamster cell lines transformed by polyoma or adenovirus type 12, under conditions which did not inhibit the growth of two non-transformed hamster cell lines [22]; this action seems to be due to the side-chain of the rifamycin derivatives, since the aminopiperazine alone had the same effect. For this reason, this mode of action will be discussed in the next section, dealing with aminopiperazine molecules. It was also observed that SV40-transformed human cells were more susceptible to growth inhibition than their non-transformed counterparts [23].

These results must be kept in mind when examining the controversial results about the effect of rifamycin derivatives on cells transformed by RNA tumour viruses: one would expect that, with these cells, the action of the drugs might result from two effects: one general effect on transformed cells, common to an effect on cells transformed by DNA viruses, and another effect on the reverse transcriptase of the RNA tumour viruses. Authors agree that rifampicin [24, 25], AF/ABDMP [25] and N-demethylrifampicin [26] cause little or no inhibition of the replication of Rous Sarcoma virus [24, 26] or of Moloney sarcoma virus [25], although they inhibit focus formation at doses of 10-50 $\mu$g/ml. But some experiments indicate that cell transformation by the virus is an event more susceptible to the drugs than is the replication of the already transformed cells [24, 25], while other experiments point to a specific effect of the drug on the multiplication of the transformed cells [26].

The rifamycin derivatives may also be active on specific cellular events other than tumour transformation. It has been reported that myeloblasts cannot differentiate into tubules in the presence of concentrations of rifamycin derivatives which allow the replication of the myeloblasts [27]. These results may be related to the description of a specific effect of AF/ABDMP on one early stage of oogenesis of the amphibian *Xenopus laevis*; in this system, amplification of the ribosomal genes occurs at a specific stage, and the synthesis of this rDNA is inhibited by 100 $\mu$g/ml AF/ABDMP, while the synthesis of chromosomal DNA is not affected [28]: these observations were confirmed by autoradiographic techniques [29]. They support, but do not prove, the hypothesis that rDNA amplification occurs through RNA-dependent DNA synthesis [30].

It had been suggested, on the other hand, that similarities might exist between gene amplification and the replication of episomes in bacteria. It may be significant that the latter phenomenon has now been shown to be susceptible to AF/ABDMP, but not to rifamycin, indicating that the effect is distinct from an inhibition of the bacterial transcriptase [31].

N-Demethylrifampicin and AF/ABDMP thus appear as drugs capable of distinguishing between cell replication on the one hand, and differentiation or tumour transformation on the other hand.

## ACTION OF AMINOPIPERAZINES

Early results suggested that the hydrazone side chain of rifamycin derivatives is essential for inhibition of vaccinia virus replication [32] but it has now been proved that an identical block of virus maturation can be obtained with rifamycin derivatives which do not contain the hydrazone chain [13, 16].

On the other hand, the molecule 1-amino-4-methylpiperazine, which is the side chain of rifampicin, does not block virus maturation and was found devoid of activity on vaccinia virus growth in the hamster BHK cells [33]. We had observed, however, that this aminopiperazine inhibited plaque formation by vaccinia virus in chick embryo fibroblasts [34], and our results prompted the preparation of a series of 1-aminopiperazines which were tested for their activity on the growth of various viruses [35]. The compounds AP4, AP5, AP8, AP22 (see Fig. 1) inhibit the formation of vaccinia virus plaques by 90% at doses of 2 to 50 μg/ml and, at two to five times these doses, they display this activity on viruses of the herpes group, but not on several types of RNA viruses. The example of Fig. 2 was selected to show that aminopiperazine number 22 (AP/22) decreased both the number and the size of the plaques formed by one herpes virus (herpes simplex type 2) with no effect on plaque formation by

**Table 1.** Comparison of various activities of ansamycin antibiotics.

| | Minimal active concentration (in μg/ml) on | | | |
|---|---|---|---|---|
| Antibiotic | Cells | Virus polymerase | Virus maturation | Virus growth |
| Rifamycin SV | 100 | 200 | 1000 (> 200) | 100-200 |
| 3-formylrifamycin SV | 100 (50) | 150 | 200 (100) | 50 |
| Compound 39 | > 100 | 200 | 500 | > 100 |
| 18 | 100 | 200 | 100 | 50 |
| N-Demethylrifampicin | > 100 | > 200 | 200 (100) | > 100 |
| Rifampicin | > 100 | > 200 | 100 (50) | 50-100 |
| AF/AB DMP | 50 | 150 | > 1000 | > 10 |
| Tolypomycin | 50 | 200 | 300 | > 10 |
| Streptovaricin | > 100 | > 200 | 1000 | > 100 |
| AF/DMI | (50) | | (25) | (25-50) |

* This table summarizes the results obtained by two groups of authors. The figures between brackets were obtained with Hela cells, and maturation block was measured by counting envelope precursors which accumulate during interruption of virus assembly [16]. The other figures were obtained with BHK21 hamster cells and the action on virus maturation was evaluated by studying the ability of ansamycin antibiotics to maintain the maturation block after removal of rifampicin [13, 14].

Newcastle Disease virus (NDV), a paramyxovirus, indicating that the action of AP/22 is not due to some non-specific effect on cells. A paradoxical increase in the number of plaques was even observed when a strain of influenza virus (WSN) was grown in the presence of AP/22; a possible explanation for this effect will be suggested later on. Several viruses of the herpes group are susceptible to the action of aminopiperazines; in particular, the action of aminopiperazine 4 (AP4) was assayed on a Burkitt cell line which contains the Epstein-Barr virus, as detected by a specific immunofluorescent stainable antigen. Considerably fewer cells contained the antigen after growth in the presence of AP/22 [36].

Several other drugs are already known to inhibit the growth of herpes viruses. These are inhibitors of DNA synthesis, such as 5-iododeoxyuridine and cytosine arabinoside, and, very recently [37], an inhibitor of protein synthesis:

asparaginase. We tested the combined action of each of these drugs with that of AP/22 and found additive and even synergistic effects (Fig. 3).

In attempts to elucidate the mode of action of the drugs, it was first observed that mutants of vaccinia virus, which have been selected for resistance to rifampicin, were still susceptible to the action of the aminopiperazines, indicating that the target sites for the effect of the two types of drugs are different. No mutants resistant to aminopiperazines could be obtained after prolonged growth of vaccinia and herpes viruses in the presence of various concentrations of the drugs. It was, on the other hand, observed that chick embryo fibroblasts treated for 24 hr or 48 hr with 100-50 $\mu$g of AP/4 or AP/22

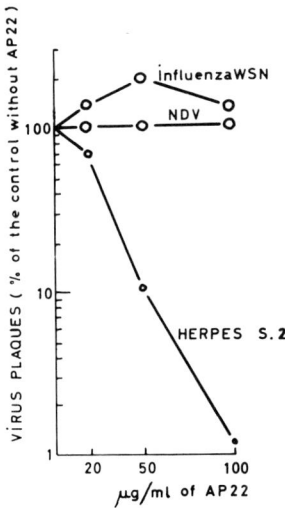

**Figure 2.** Number and size of plaques of three viruses in the presence of various concentrations of AP/22 in agar or methylcellulose overlay (○ normal, large plaques; ○ medium size plaques; ● small plaques). NDV: Newcastle Disease Virus, a RNA virus of the paramyxovirus group; WSN: a strain of influenza virus of the myxovirus group; Herpes Simplex type 2, a DNA virus of the herpes group.

were partly resistant to herpes simplex virus, even if the drug was not present during virus growth. These two facts pointed to the possibility that inhibition of virus plaque formation by aminopiperazines may be mediated by an action on some cellular function which is more essential for the synthesis of herpes and vaccinia viruses than for the synthesis of several RNA viruses. Aminopiperazine did not modify uridine and amino acid incorporation into acid insoluble components of uninfected cells, but there was an effect on thymidine incorporation (Fig. 4). This incorporation was increased by 15-22% after cells had been grown for 2 hr in he presence of 50-100 $\mu$g/ml of AP/22 or for 20 hr in the presence of 50 $\mu$g/ml of the drug; however, a prolonged treatment of 20 hr with the high dose of drug (100 $\mu$g/ml) led to more than 50% inhibition of thymidine incorporation. These effects might be due to modifications of the thymidine pool by the drug or to inhibition of some type of DNA synthesis accompanied by increase of a repair mechanism; within the frame of the second hypothesis, it would appear, however, that increased DNA synthesis is the primary event, followed by inhibition, since cells treated with 100 μg/ml of AP/22 for 2 hr and 20 hr first showed an increase and, later, a decrease of thymidine incorporation. The hypothesis of the induction of a repair mechanism by AP/22 is sustained by two experiments. First, the previous experiment was repeated but [$^3$H]-thymidine was incorporated for 2 hr either alone or in the presence of 1 mM hydroxyurea. Under the latter conditions, it has been claimed that semi-conservative DNA synthesis is selectively abolished and that the remaining thymidine incorporation measures unscheduled DNA synthesis, in

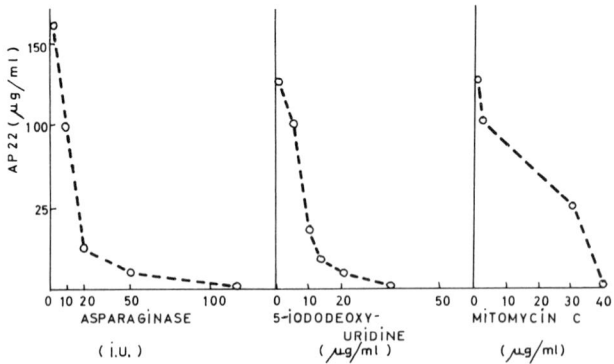

Figure 3. Inhibition of herpes simplex virus plaques by AP/22, assayed in combination with asparaginase, 5-iododeoxyuridine or mitomycin C, the pair of drugs being present in the methylcellulose overlay. Each point in the graphs corresponds to the combined dose which caused 90% inhibition of plaque formation.

Hela cells [38]. Apparently this type of synthesis was increased by AP/22, since thymidine incorporation in the presence of hydroxyurea represented a higher percentage of total incorporation in cells treated with AP/22 than in non-treated cells (Fig. 5), and the percentage was highest under conditions where total incorporation was decreased by a treatment with 100 μg/ml of AP/22 for 20 hr. In a second type of experiment, AP/22 treated cells were irradiated with uv light, in order to see whether the repair mechanism induced by uv was modified by the drug. Monolayers of chick embryo fibroblasts were irradiated with uv light and treated with 100 μg/ml of AP/22 either for 24 hr before irradiation or for 4 hr after. In all cases, [$^3$H]-thymidine was incorporated 2 to 4 hr after irradiation, with or without hydroxyurea added (Fig. 6). The results first substantiate the assumption that thymidine incorporation, measured in the

# ACTION OF RIFAMYCIN AND AMINOPIPERAZINE DERIVATIVES

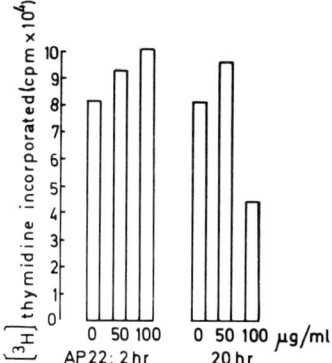

**Figure 4.** Action of AP/22 on [$^3$H]-thymidine incorporation into acid-insoluble material of chick embryo fibroblasts. 2.5 μCi/ml of [$^3$H]-thymidine were added to the nutrient medium either during the 2 hr treatment with AP/22 or during the last 2 hr of a 20 hr treatment with the drug. The reaction was stopped with ice-cold 0.5 N perchloric acid, the precipitate was washed three times with this solution and the pellet was finally solubilized in toluene for 30 min at 60°C, and then added to 1 ml alcohol and 10 ml scintillating fluid.

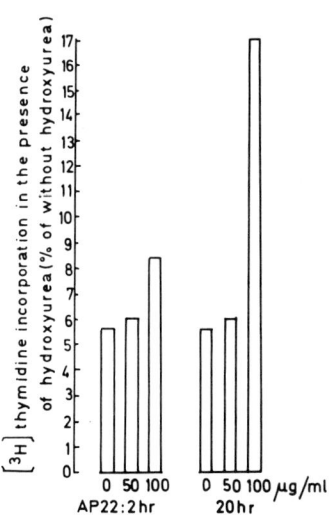

**Figure 5.** [$^3$H]-Thymidine incorporation in the presence of 1 mM hydorxyurea under conditions otherwise similar to those described in Fig. 4. Two assays of thymidine incorporation were run in parallel, with or without hydroxyurea. Results are expressed in percentages of the cpm found for thymidine incorporation in the absence of hydroxyurea.

presence of hydroxyurea, mainly represents DNA synthesis of the repair type, since residual thymidine incorporation was the same after uv or hydroxyurea treatment. As regards total thymidine incorporation in the absence of hydroxyurea, this was decreased by 50% after uv irradiation, but the decrease was only about 29% when the cells were treated with AP/22 for 4 hr after uv. Actually uv light induced even a slight increase of thymidine incorporation when irradiation was performed on cells pretreated for 24 hr with AP/22.

To exclude the possibility that the results were due to modifications of thymidine pools, the most direct approach was to test whether *in vitro* DNA synthesis was modified when nuclear extracts were prepared from AP/22 treated

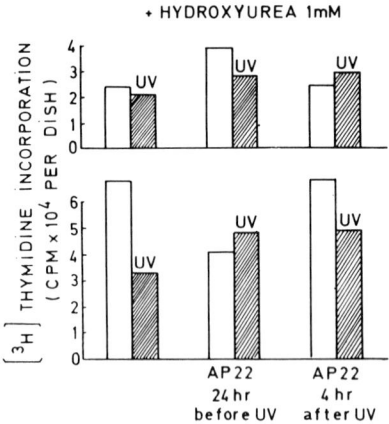

Figure 6. Influence of AP/22 on the repair DNA synthesis induced by uv irradiation. Chick embryo fibroblasts were irradiated with 200 ergs/mm$^2$ of uv light, either after 20 hr treatment with AP/22, or immediately before a 4 hr treatment with this drug. [$^3$H]-thymidine was added in each case 2 to 4 hr after uv irradiation.

cells. Nuclear extracts were assayed for [$^3$H]-thymidine triphosphate (dTTP) incorporation into acid insoluble material in the presence of the other three deoxyribonucleotides. The endogenous DNA synthesis, as assayed in the absence of added DNA, was very low under our conditions. With exogenous DNA added, dTTP was incorporated linearly with time, and this incorporation was not inhibited by the addition of 100 μg/ml AP/22 to the assay mixture.

We then investigated whether *in vivo* treatment of the cells would influence the amount of DNA polymerase available, as tested with the *in vitro* assay in the presence of exogenous DNA. Nuclear extracts were prepared from cells pretreated *in vivo* with 100 μg/ml AP/22 for 4 hr at 37°C, and assayed for DNA polymerase activity *in vitro*. The rate of [$^3$H]-dTTP incorporation was greater with extracts obtained from AP/22-treated cells than with those from untreated cells (examples of such results will be shown below).

How could this increase of DNA polymerase activity produce an inhibition of virus growth? It has been shown that viruses of the herpes group completely stop cellular DNA synthesis a few hours after they have entered the cells [39, 40]. The mechanism of this switch-off has not been elucidated, but one can readily see that it could be useful for replication of the viral DNA if it enlarges the pool of available deoxyribonucleotides. Experiments are now in progress to determine whether herpes viruses lose their ability to switch off cellular DNA synthesis when they infect AP/22 treated cells; if this were so, cellular DNA would then compete with the virus for the utilization of the pool of macromolecules, and this would decrease the yield of virus per cell. This hypothesis fits well with the fact that aminopiperazines have been shown to increase the number of plaques formed by a strain of influenza virus. Viruses of this group depend on the activity of cellular DNA for their growth, although it is not clear whether it is cellular DNA synthesis, or transcription, which is needed for the synthesis of influenza viruses to occur. Cells pretreated with low doses of actinomycin D or uv light will sustain the growth of several types of viruses, but not that of influenza viruses. Since aminopiperazines were found to increase mainly DNA synthesis of the repair type, it is conceivable that this increase may facilitate the replication of the virus.

Another possibility is that the mild action of AP/22 on DNA synthesis of the cells has no correlation with its antiviral effect. Other consequences of the treatment with AP/22 should be looked for.

Long-term effects of aminopiperazines on the growth of continuous cell lines were then studied [22]. Two hamster cell lines ($BHK_{21}$ and Nil) could be cultivated and subcultivated in the continuous presence of 50 $\mu$g/ml of AP/8 or $N$-demethylrifampicin, but two transformed cell lines died after one or several subcultures; these were the two previously cited hamster cell lines which had been transformed by two different oncogenic viruses: polyoma and adenovirus type 12. The toxic effect on transformed cells is a long-term one since the number of cells increases normally during the first four days of drug treatment. In addition the amount of drug required to obtain the toxic effect is rather high, since the nutrient medium was replaced daily with fresh medium plus drug, and since aminopiperazines are probably not metabolized in the cells.

In order to determine whether this long-term toxic effect on transformed cells was related to exaggeration of the previously described action on DNA synthesis and increased polymerase activity, the experiments performed with chick embryo fibroblasts were repeated with transformed and untransformed hamster cells. No clearcut differences between these two types of cells could be found, as regards DNA synthesis in response to aminopiperazine treatment. One experiment is shown in Fig. 7. HTr7 cells transformed by adenovirus type 12, and untransformed BHK 21 cells, were treated for 15 hr with 100 $\mu$g/ml AP/22; nuclear extracts were then prepared and assayed for DNA polymerase activity in the presence of native or denatured thymus DNA. DNA polymerase activity of AP/22-treated cells was greater than that of untreated cells; with HTr 7 extracts, this increased activity was found mainly when using denatured DNA as template, which is in keeping with the hypothesis that the increased polymerase is a repair enzyme; however, this specificity was not found with untransformed hamster or

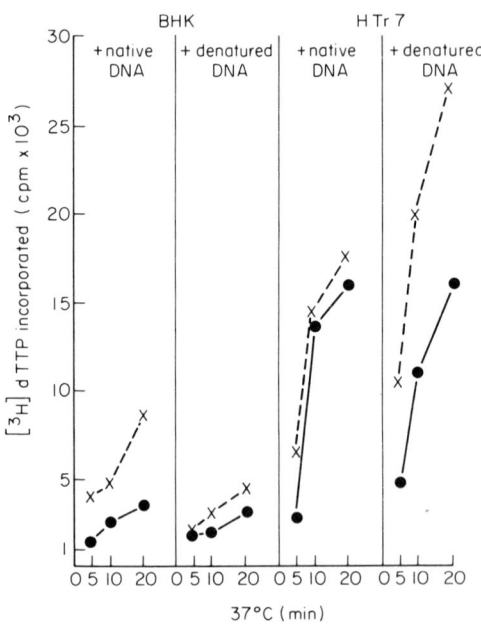

Figure 7. *In vitro* assay of DNA polymerase activity of nuclei extracts from hamster cells, either transformed (HTr7) or non-transformed (BHK21), which had been treated for 15 hr at 37°C with 100 μg/ml AP/22 (x - - - - x), or had not been treated (●————●). For each reaction mixture, nuclei were prepared from $6 \times 10^6$ cells, after swelling for 10 min in cold hypotonic buffer (0.01 M Tris-HCl pH 8.6, 0.01 M KCl, $1.5 \times 10^{-3}$ M $MgCl_2$), followed by addition of 0.1% Nonidet. Cells were broken with three strokes of a Dounce homogenizer and centrifuged at 900 $g$ for 15 min. The pellet was resuspended in 50 μl of 0.1 M phosphate buffer pH 7.2, which was then added to a reaction mixture containing, in 230 μl, 0.02 μmole of dATP, dGTP, dCTP each, 0.015 μmole ATP, 10 mM dithiothreitol, 40 mM potassium phosphate, 8 mM $MgCl_2$, 15 μg of thymus DNA. The DNA was native or denatured by boiling and quick cooling. The reaction was stopped by addition of 10 ml 10% trichloracetic acid with 1% pyrophosphate plus 2 mg albumin and the precipitate, washed in this solution, was finally solubilized as described under Fig. 4.

chick cells, and it is not yet clear whether it is a characteristic of transformed cells. For the moment, there is no obvious correlation between toxic effects of aminopiperazines and the extent of their activity on DNA synthesis.

In contrast, other experiments unexpectedly suggested correlations between the toxic effect and some morphological lesions produced by aminopiperazines in the cells. These lesions were discovered when investigating the action of aminopiperazines on nervous tissue infected with herpes simplex virus, an experiment which was prompted by the fact that this virus sometimes causes severe necrotizing encephalitis in man.

Organotypic nervous tissue cultures were prepared from the brain and ganglia of mice: in this system, it is not possible to assay the toxic effect of drugs by following the multiplication of cells or testing their ability to be subcultured. Light and electron microscope studies were made [41]. Two or three days after addition of 50 μg/ml AP/22, all types of cells had retained normal structures with normal aspect of membranes but atypic formations appeared in the cytoplasm of neurones; they consisted of lamellar formations which seemed to be located in the lysosomes: they were osmiophilic and probably of lipidic nature. Preliminary results indicate that, in cultures inoculated with $5 \times 10^4$ PFU of herpes simplex virus type 2 in 0.05 ml nutrient medium, the drug decreased the production of mature virions with no accumulation of immature particles. The lamellar formations were also seen in drug-treated hamster cell lines, but they were much larger and more numerous in transformed cells than in normal hamster cells and in the nervous tissue of mice.

These electron microscope studies drew attention to the metabolism of lipids in drug-treated cells. The hamster cell line transformed by adenovirus type 12 was treated daily with 50 μg/ml of AP/22 and subcultivated in the presence of this drug on the fourth day. The next day, the total amount of lipids in these cells was eight times the amount found in the same cultures not treated with drug [42]. $[^{14}C]$-choline incorporation into acid-insoluble cellular material was modified as early as 20 hr after addition of one dose of 50 μg/ml AP/22, although the cells continued to multiply normally for several days despite the daily addition of 50 μg/ml AP/22. The modification of $[^{14}C]$-choline incorporation was not paralleled by a modification of $[^{14}C]$-glucosamine incorporation, indicating that the alteration of metabolism is not linked with glycolipids.

It seems, at first glance, very probable that the effect on lipid metabolism of the cells is completely unrelated to that on DNA synthesis. It may, however, be noted that a recent study has shown that, in regenerating liver, initiation of DNA synthesis is mediated by a class of DNA which is associated with the inner nuclear envelope in a complex which involves lipoproteins [43]. Whether aminopiperazines modify the metabolism of these lipoproteins is certainly worthy of further study.

## SUMMARY

Inhibition of vaccinia virus growth is produced by several rifamycin derivatives and this inhibition is better correlated with some action of the drugs on maturation of the virion than on its DNA-dependent RNA polymerase.

Aminopiperazines inhibit the growth of vaccinia and herpes viruses in a different way, which may be mediated by an action on the cells.

Rifamycin derivatives, and not aminopiperazines, specifically inhibit one or several polymerases (DNA-dependent RNA or DNA polymerases; RNA-dependent DNA polymerase).

Toxic effects of rifamycin derivatives on cells are related to either inhibition of DNA-dependent DNA polymerase or mitochondrial lesions; toxicity of aminopiperazines is related to modifications of lipid metabolism and lesions probably located in the lysosomes. This effect is possibly the cause of inhibition of cell growth, which was observed preferentially with transformed cells.

Some rifamycin derivatives also have an effect on cell differentiation, possibly by inhibiting an RNA-dependent DNA polymerase.

## REFERENCES

1. Heller, E., Argaman, M. and Goldblum, N. (1969). *Nature, Lond.* **222**, 273.
2. Subak-Sharpe, J. H., Timbury, M. C. and Williams, J. F. (1969). *Nature, Lond.* **222**, 341.
3. Moss, B., Rosenblum, E. N., Katz, E. and Grimley, P. M. (1969). *Nature, Lond.* **224**, 1280.
4. Grimley, P. M., Rosenblum, E. N., Mims, S. J. and Moss, R. (1970). *J. Virol.* **6**, 519.
5. Nagayama, A. B., Pogo, G. T. and Dales, S. (1970). *Virology* **40**, 1039.
6. Pennington, T. H., Follett, E. A. C. and Szilagyi, J. F. (1970). *J. gen. Virol.* **9**, 225.
7. Moss, B., Rosenblum, E. N. and Grimley, P. M. (1971). *Virology* **45**, 123.
8. Moss, B., Rosenblum, E. N. and Grimley, P. M. (1971). *Virology* **45**, 135.
9. Katz, E. and Moss, B. (1970). *Proc. natn. Acad. Sci. U.S.A.* **66**, 677.
10. Katz, E. and Moss, B. (1970). *J. Virol.* **6**, 717.
11. McAuslan, B. R. (1969). *Biochem. biophys. Res. Commun.* **37**, 289.
12. Pogo, B. G. T. (1971). *Virology* **44**, 576.
13. Pennington, T. H. and Follett, E. A. C. (1971). *J. Virol.* **7**, 821.
14. Szilagyi, J. F. and Pennington, T. H. (1971). *J. Virol.* **8**, 133.
15. Subak-Sharpe, J. H., Pennington, T. H., Szilagyi, J. F., Timbury, M. C. and Williams, J. F. *In* "RNA Polymerase and Transcription: Proceedings First Lepetit Colloq." (L. Silvestri, ed.) North Holland Publishing Co., Amsterdam.
16. Grimley, P. M. and Moss, B. (1971). *J. Virol.* **8**, 225.
17. Engelberg, H. and Soudry, E. (1971). *J. Virol., Lond.* **7**, 847.
18. Gurgo, C., Ray, R. K., Thiry, L. and Green, M. (1971). *Nature New Biology* **229**, 111.
19. Gallo, R. C., Yang, S. S. and Ting, R. C. (1970). *Nature, Lond.* **228**, 927.
20. Green, M. Communicated at the 2nd International Congress of Virology, Budapest 1971.
21. Brockman, W. W. and Carter, W. A. (1971). *Nature, Lond.* **230**, 249.
22. N. Van Tiegham. Manuscript submitted for publication.
23. A. Vaheri. Personal communication.
24. Diggelmann, H. and Weissmann, C. (1969). *Nature, Lond.* **224**, 1277.

25. Calvin, M., Joss, U. R., Hackett, A. J. and Owens, R. B. (1971). *Proc. natn. Acad. Sci. U.S.A.* **68**, 1441.
26. Vaheri, A. and Hanafusa, H. Manuscript in preparation.
27. Luzzati, D. Communicated at the 7th FEBS meeting of VARNA, 1971.
28. Crippa, M. and Tocchini-Valentini, G. P. (1971). Communicated at the 7th FEBS meeting of VARNA. *Proc. natn. Acad. Sci. U.S.A.* (In Press).
29. Ficq, A. and Brachet, J. *Proc. natn. Acad. Sci. U.S.A.* (In Press).
30. Tocchini-Valentini, G. P. and Crippa, M. (1971). *In* "2nd Lepetit Colloquium on Oncogenic Viruses," (L. Silvestri, ed.), p. 237, North Holland Publish. Amsterdam.
31. Tocchini-Valentini, G. P. Communicated at the 7th FEBS meeting of Varna, 1971.
32. Zakay-Rones, Z. and Becker, Y. (1970). *Nature, Lond.* **226**, 1162.
33. Follett, E. A. C. and Pennington, T. H. (1971). *Nature, Lond.* **230**, 117.
34. Thiry, L. and Lancini, G. (1970). *Nature, Lond.* **227**, 1048.
35. Lancini, G., Cricchio, and Thiry, L. J. (1971). *J. Antibiotics* **24**, 64.
36. Thiry, L. and Lamy, M. (1971). *In* "The Biology of Oncogenic Viruses; Second Lepetit Colloquium," (L. Silvestri, ed.), p. 244.
37. Maral, R. and Werner, G. H. (1971). *Nature New Biology, Lond.* **232**, 187.
38. Painter, R. B. and Cleaver, J. E. (1967). *Nature, Lond.* **216**, 369.
39. Kaplan, A. S. and Ben-Porat, T. (1963). *Virology* **19**, 205.
40. Ben Porat, T. and Kaplan, A. S. (1965). *Virology* **25**, 22.
41. M. Dubois-Dalcq. Personal communication.
42. J. de Bruyn. Personal communication.
43. Mizuno, N. S., Stoops, C. E. and Peiffer, R. L. (1971). *J. molec. Biol.* **59**, 517.

## DISCUSSION

**U. Bachrach:** Is there any effect of aminopiperazine on *E. coli* or on phage-infected cells?

**L. Thiry:** Aminopiperazines exhibit no activity against most of the known polymerases (transcriptases, reverse transcriptases, DNA polymerases) so that this differentiates them from the rifamycin derivatives. I do not know whether the aminopiperazines have been tested on phage-infected bacteria, but they are not antibacterial agents.

**U. Bachrach:** What is known about the therapeutic value of aminopiperazines (in virus-infected animals).

**L. Thiry:** I have been informed that $AP_4$ decreases the number of poxes in the tail of mice innoculated with vaccinia virus.

**B. Rada:** In the enhancement of WSN influenza virus, was there an increase only in the number of plaques, or did the size of the plaques also increase? Or did you test the enhancement in another system where the virus yield was estimated? In the event of an increase only in plaque count, this might indicate that your compound is also involved in adsorption or penetration of the virus.

**L. Thiry:** Plaque size was not increased. However the possibility of some influence of the compound on virus adsorption or penetration cannot be ruled out.

**B. Rada:** This would fit in with the second mode of action of your compound, i.e. on lipid metabolism and, by inference, on the cell membrane.

**W. H. Prusoff:** Does the macrocyclic moiety of rifamycin have to be cleaved from the piperazine moiety in order to obtain antiviral activity?

**L. Thiry:** I'm afraid I do not know.

**U. Bachrach:** The work of some colleagues at the Department of Virology of the Hebrew University (Jerusalem) indicated that pre-incubation enhanced the antiviral activity of rifampicin, thus pointing to the possibility that it is metabolized.

**O. P. van Diggelen:** Are specific activities of DNA, and pool sizes for thymidine and its phosphorylated derivatives, unchanged after AP22 treatment of cells? A small change in pool sizes could explain the 20-30% difference observed for thymidine incorporation, could it not?

**L. Thiry:** Aminopiperazines did not modify the amount of labelled thymidine in the acid-soluble extract. The drugs also provoked an increase in DNA polymerase activity, when tested *in vitro* with added exogenous DNA.

# Alkylated Pyrimidine Nucleosides and (Poly)nucleotides as Potential Anti-viral Agents

## D. SHUGAR

*Institute of Biochemistry and Biophysics, Academy of Sciences; and Department of Biophysics, University of Warsaw, Warsaw*

The problems involved in the development of suitable anti-viral (and anti-tumour) agents have been extensively formulated and reviewed by a number of investigators, including one of the speakers at this Symposium [1, 2, 3, 4]. The principal aim of this paper is to draw attention to some pyrimidine nucleoside, nucleotide and polynucleotide analogues, with particular emphasis on those alkylated either on the aglycone or/and on the sugar moiety, the potentialities of which have hitherto by no means been fully exploited for the development of new anti-metabolites or for use as model systems in studies on the mode of action of some known anti-viral or anti-tumour agents.

## PYRIMIDINE DERIVATIVES

The literature abounds with reports of anti-viral activities of a multitude of pyrimidine analogues. For example, 5-alkyl and 5-aryl substituted 6-pyrimidine carboxylic acids have been claimed to exhibit activity against herpex simplex and influenza viruses [5]. Ring $N_1$ substituted orotic acids [6] and ring $N_1$ alkyl (or aryl) substituted 6-aminouracils [7] reportedly possess therapeutic activity against a number of viruses; while 3-allyl-5-alkyl uracils have been found to inhibit herpes and vaccinia viruses in tissue culture and experimental herpes keratitis in the rabbit [8]. Gauri and Rhode [8] claim their derivatives to be effective in the treatment of all types of recurring herpetic skin and mucous diseases. A compilation of additional pyrimidine derivatives with reported anti-viral activity may be found in two reviews (Cheng, [9], Cheng and Roth, [10]).

Earlier studies by Muraoka, *et al.* [11] on the potential anti-viral activity of 5-alkyluracils, 5-alkylisocytosines and 5-alkyl-6-methylisocytosines, and which

---

The following abbreviations are used: EtUdR, 5-ethyldeoxyuridine; FUdR, 5-fluorodeoxyuridine; IUdR, 5-iododeoxyuridine; BUdR, 5-bromodeoxyuridine; ara-C, arabinofuranosylcytosine.

demonstrated some activity of 5-hexyl-6-methylisocytosine against type-1 strain of adenovirus, but not against polio or vaccinia viruses, were subsequently extended by the same authors [12] to a broader series of analogues and a variety of RNA and DNA viruses. Pronounced inhibitory effects against both RNA and DNA viruses were reported for some of these derivatives, particularly 5-butyluracil, the potency being somewhat enhanced when 5-butyluracil was replaced by 5-butyluridine; the latter was claimed to exhibit a wider spectrum of action against DNA viruses relative to 5-fluorodeoxyuridine, which was used as a control [12]. These compounds were found to inhibit only the early phase of viral reproduction [13]. It is, perhaps, of interest in this connection that, whereas 5-ethyluracil can partially replace thymine in the DNA of thymineless mutants of *E. coli* and lead to thymineless death of the latter, 5-butyluracil and 5-hexyluracil were quite inert in this system [14].

A well authenticated example of a free pyrimidine as an effective anti-viral agent is 2-thiouracil, which is known to inhibit a number of plant viruses [15], presumably as a result of its ready incorporation into viral RNA. But, in general, reports of activity of this agent in mammalian systems have not been such as to inspire confidence. It has recently been reported that 2-thiouridine inhibits multiplication of Coxsackie A-21 virus in *in vitro* systems, presumably by inactivation of the viral RNA following incorporation of the analogue [16], but no direct evidence to this effect was proffered.

Another example is 6-uracil methylsulphone, which was found to inhibit infective herpes simplex *in vitro* by some mechanism other than direct inactivation of the virus [17].

On the whole, notwithstanding the voluminous literature on the subject, free pyrimidine derivatives have not been extensively employed as potential anti-viral or anti-tumour agents. In most instances where such activity has been claimed, there is a scarcity of data regarding the possible mode of action. In part, at least, this is due to the fact that pyrimidines are relatively little utilized for incorporation into nucleic acids by mammalian systems. A notable exception is 5-fluorouracil, undoubtedly the most intensely investigated of potential pyrimidine anti-metabolites; this is readily converted to the riboside in both bacterial and mammalian systems. Subsequent enzymatic phosphorylation is followed by extensive incorporation into the RNA of mammalian cells. But, while it is active against DNA viruses in cell cultures, its efficacy *in vivo* is negligible. This subject has been extensively reviewed by Heidelberger [18, 19].

It is rather surprising that the biochemistry of 5-alkyluracils has been so little investigated. It was shown by Barrett *et al.* [20] that such compounds are effective inhibitors of dihydrouracil dehydrogenase from rat tissue supernatants, thus preventing the degradation of both uracil and thymine. In a bacterial system, 5-ethyluracil was found to be a relatively poor substrate for thymidine phosphorylase [14], and this was regarded as one of the factors accounting for the only 18% replacement of thymine by 5-ethyluracil in thymineless mutants of *E. coli.*

## NUCLEOSIDE ANALOGUES

Even in those instances where a pyrimidine derivative appears to provide a block to viral growth, it is usually the corresponding deoxyriboside which is the more effective. Probably the most striking illustration is IUdR, active against herpes simplex and vaccinia viruses, and the first purely synthetic compound to be employed clinically as an anti-viral agent, viz. against herpes simplex infections of the corneal epithelium in man [21, 22]. The mode of action of this anti-metabolite is still the subject of some controversy, the bulk of the evidence suggesting that the primary site of viral inhibition is that following its incorporation into viral DNA in place of thymidine. This subject has been treated by a more competent author, (see p. 135), and no attempt will be made to do so here.

It is not at all surprising that the success, albeit modest, achieved with IUdR should have stimulated widespread trials with other pyrimidine deoxyribonucleoside analogues and the synthesis of new ones. These may be divided into two classes: (a) those in which the aglycone has been modified by substitution at the 5- or/and, less frequently, the 6-position; (b) those in which the carbohydrate moiety has been substituted or replaced by another sugar such as D-arabinose.

A typical example is 5-methylamino-2′-deoxyuridine, which, although only a weak thymidine antagonist in bacterial systems [23], compares favourably with IUdR in activity against herpes simplex both in tissue culture and in rabbit eye keratitis systems [24]. Particularly interesting is the high degree of selectivity of this analogue, it being apparently without effect on other viral systems tested.

Following the demonstration of the clinical utility of IUdR, it was quickly shown that equal efficacy was exhibited by BUdR [25], and even more so by 5-trifluoromethyldeoxyuridine [26]. Other iodinated pyrimidine nucleosides have also been found effective (see review by Prusoff [7]).

The mode of action of various analogues is, however, not necessarily identical, as already clear from the high selectivity of 5-methylamino-2′-deoxyuridine, referred to above. A striking example of this is the work of Walker *et al.* [27], using an *in vitro* method for testing the ratio of tissue tolerance to viral susceptibility. It was found that IUdR was more effective than FUdR against type 1 herpes simplex virus, whereas the reverse was true for type 2. By contrast, ara-C inhibited both types equally, whereas none of the three compounds exhibited activity against adenovirus.

## 5(6)-ALKYLPYRIMIDINE NUCLEOSIDES

The chemotherapeutic applications of IUdR and related analogues have been limited largely to topical treatment, due to their known ability to undergo incorporation into the DNA with its attendant risk of mutagenic action. While incorporation into DNA (or RNA) is not necessarily the mechanism by means of which a viral anti-metabolite operates, nonetheless one of the criteria advanced as essential for maximal efficacity of a viral or tumour anti-metabolic agent is lack of incorporation into cellular nucleic acids. This is regarded as synonymous with the absence of any mutagenic effects.

Although this undoubtedly represents the ideal objective, no consideration appears to have been given to the possibility of a base analogue which might exhibit the same base-pairing properties as the parent base, so that its incorporation could conceivably be non-mutagenic. This point first came to our attention during the course of an investigation on the properties of poly 5-ethyluridylic acid, carried out primarily with a view to clarifying the role of thymine in DNA as compared to uracil in RNA [28, 29].

Contrary to other base analogues, the pK for dissociation of the ring $N_3$ hydrogen in 5-ethyluracil and its ribosides was found to be practically identical with the corresponding values for thymine and its glycosides [30], as might have been anticipated. This is of some importance since the ring $N_3$ hydrogen is the donor in Watson-Crick base-pairing.

It was soon shown by Dr. M. Piechowska that 5-ethyluracil, an obvious thymine analogue, could replace thymine to an appreciable extent (up to 18%) in thymineless mutants of *Escherichia coli* [14]. This, in turn, prompted us to synthesize 5-ethyldeoxyuridine [31], which was shown by Dr. I. Pietrzykowska to undergo incorporation into phage DNA, to the extent of almost 70% replacement of thymidine in phage $T_3$ DNA, measured by actual isolation of the latter and hydrolysis and chromatography of the bases [32, 33]. It was, in fact, this ability to support growth of thymineless mutants of *E. coli*, and of phage $T_3$ (as well as other T phages), which made it possible initially to distinguish the α- and β- anomers of EtUdR, later confirmed by NMR spectroscopy [31]. Subsequently the synthesis of 6-[$^3$H]-5-ethyldeoxyuridine was described by Gauri *et al.* [34]. More recently, the nucleoside has been prepared with a [$^{14}$C]-label in the ethyl side chain (see below).

The possible mutagenic activity of EtUdR was examined by Dr. I. Pietrzykowska, with BUdR as control, by means of the spot test of Freese [35], using 16 rII mutants of phage $T_4$ and looking for reversions to $r^+$. Under conditions where BUdR gave the expected frequencies of reversions, EtUdR was quite without effect [28, 31]. This result was subsequently confirmed by Prof. C. Heidelberger in Madison. Reported lack of mutagenicity of EtUdR in a higher organism, *Drosophila* [36], must be treated with reserve, since no evidence for incorporation into cellular DNA was demonstrated.

The antiviral activity of EtUdR was examined by Prof. C. Heidelberger against vaccinia virus on HeLa cells. The α-anomer, as expected, was inactive: whereas the β-anomer was nearly as effective as IUdR and BUdR [31]. The activity of the analogue has also been tested in the laboratory of Prof. H. E. Kaufman against herpes simplex virus by the plaque reduction method on chick embryo fibroblast monolayers. Inhibition of viral growth was observed at all dilutions, with IUdR as control, but the activity was somewhat less than that of the control [37]. While it is unlikely, as pointed out by Kaufman [37], that EtUdR could profitably supersede such anti-metabolites as IUdR or trifluoromethylthymidine for *topical* use, it may nonetheless prove superior with deeper infections because of the toxicity and mutagenicity of the latter. A more extensive testing program is consequently envisaged, but requiring the large-scale synthesis of EtUdR. Meanwhile the activity of this analogue against herpes simplex and vaccinia viruses has also been reported by Gauri *et al.* [38]. However, no investigations have as yet been made as to the possible mode of action of EtUdR as an anti-viral agent.

It would obviously be of value to synthesize and test 5-alkyldeoxynucleosides with higher alkyl side chains. It should be recalled (see above) that the *ribo*side of 5-butyluracil was reported to exhibit a wide spectrum of anti-viral activity [12], although independent confirmation is desirable.

The ribonucleosides of 6-methyluracil and 6-methylcytosine have both been found moderately active in *in vitro* assays against herpes simplex [39]. No attempt has yet been made to examine the biochemical basis of this inhibition, but it is almost certainly not without relevance that these nucleosides exist in the *syn*, as compared to the normal *anti*, conformation. The activities of the corresponding deoxynucleosides, i.e. 6-methyldeoxyuridine and 6-methyldeoxycytidine have not as yet been reported, but preliminary trials were apparently not encouraging [40]. The procedures hitherto employed for the synthesis of these analogues, as well as the corresponding 5,6-dialkylnucleosides, have been based on the application of standard condensation techniques [41, 42]. The relatively low yields obtained prompted Otter *et al.* [43] to undertake the synthesis of 6-carbon-substituted pyrimidine nucleosides *via* Claisen rearrangements from 5-hydroxypyrimidine nucleosides; the resulting successful synthesis of 5-allyloxyuridine, and its Claissen rearrangement to provide 6-allyl-5-hydroxyuridine in good yield undoubtedly warrants an examination of the generality of this technique.

But an even more promising, and more general, approach to the synthesis of both 5- and 6-alkyl, and 5,6-dialkyl, pyrimidine nucleosides, may be found in a procedure hitherto described only in preliminary form by Pichat *et al.* [44]. This is based on relatively simple modifications of the laborious *n*-butyllithium technique of Ulbricht [45], which has not found many proponents because of its extremely low yields. By trimethylsilylation of the sugar hydroxyls of ribo- or deoxyribo- nucleosides, and addition of hexamethylphosphortriamide immediately after the lithiation step, Pichat et al. [44] obtained 54-87% yields of mixtures of 5-methyl and 6-methyl nucleosides, with only traces of the $N_3$-methyl derivatives. The method has been applied to the preparation of [$^{14}$C]-5-methyl-labelled nucleosides from uridine (or 5-bromouridine) and deoxyuridine (or 5-bromodeoxyuridine). A sample of [$^{14}$C]-labelled EtUdR, kindly supplied to us by Dr. L. Pichat, presumably was prepared by the foregoing procedure.

The anti-viral activity data hitherto obtained with EtUdR were sufficiently encouraging to prompt undertaking the synthesis of 5-ethyldeoxycytidine. This has now been achieved by two independent routes, and is described in full detail elsewhere [46]. The possible utility of this analogue is exemplified by the application by Woodman [47] of a polyethyleneimine complex of 5-iododeoxycytidine-5'-phosphate for subcutaneous injection into mice bearing spontaneous mammary tumours. This resulted in enhanced incorporation and tumour localization of 5-iododeoxyuridine (presumably by cellular deamination), the inference being that the saltlike complex was more readily transported through the tumour cell membrane system. If, as suggested by Goz and Prusoff [2], equally enhanced uptake may occur with viral-infected cells, then 5-ethyldeoxycytidine-5'phosphate coupled with polyethyleneimine may also prove to be a useful viral inhibitor. An additional incentive for conducting tests with 5-ethyldeoxycytidine is provided by the report of Sidwell et al. [48], who studied the activity in an *in vitro* system against human cytomegalovirus of a series of 5-halogeno pyrimidine nucleosides. The deoxycytidine derivatives were found to be more active than the corresponding deoxyuridines. It should be noted that the preparation of the corresponding 5-ethylcytidine has already been reported [49]; while this may prove less useful than the deoxynucleoside, it could have other important applications following conversion to the 5'-pyrophosphate, a potential substrate for polynucleotide phosphorylase (see below).

## NUCLEOSIDES WITH MODIFIED PENTOSE RINGS

During the past four to five years considerable effort has been devoted to the synthesis of purine and pyrimidine nucleoside analogues in which the pentose moiety has been modified. Various considerations have dictated the nature of the modification introduced, e.g. substitution of the 3'- and/or 5'-hydroxyls with groups non-cleavable *in vivo* will clearly prevent incorporation into nucleic acids. Various other types of substitution may affect not only the antimetabolic activity of the parent compound, but other factors such as permeability through cell membranes, toxicity, etc. A recent review on this subject, with special emphasis on the synthetic methods employed, is that of Shen [50].

Undoubtedly the D-arabinofuranosyl nucleosides (and nucleotides) are the best-known members of this class of anti-viral and anti-tumour agents. Cytosine arabinoside is significantly inhibitory to DNA viruses and several RNA viruses, including Rous sarcoma. Its efficacy against vaccinia and herpes simplex equals that of IUdR. It is, in fact, effective against herpes simplex strains which are resistant to IUdR (see above). But its high toxicity is at variance with its anti-viral activity. Furthermore it is subject to deamination *in vivo* to the relatively inactive arabinofuranosyl uracil. In fact the efficacy of ara-C has been linked to the level of cytidine deaminase in the bone marrow of leukaemic patients [51]. The biochemistry of these compounds has been extensively reviewed by Cohen [52], and their anti-viral and anti-tumour properties by Prusoff [1], Sartorelli and Creasy [53], and Roy-Burman [4].

## NUCLEOSIDES WITH ALKYLATED SUGAR MOIETIES

In view of the role of the $2'$-hydroxyl in determining the differences in structure and function between DNA and RNA, it is rather odd that not until very recently were any attempts made to examine the possible modifications of biological activity of nucleosides as a result of alkylation of the carbohydrate moieties.

A number of nucleosides with branched-chain sugars have been synthesized by Walton *et al.* [54], and by Harper and Hampton [55]. These include, amongst others, $2'$- and $3'$-C-methyladenosines. The reasoning behind the preparation of these compounds was the proposal that the therapeutic activity of some adenosine analogues is limited by their ready conversion to the less biologically active inosine analogues through the action of adenosine deaminase. The methylated adenosines were, in fact, found to be more resistant to enzymatic deamination. They also inhibited the growth of KB cells in culture and proved to be effective anti-vaccinia agents in mice. Extension of the synthetic procedures to the preparation of $2'$- and $3'$-$C$ methylated uridine and cytidine nucleosides gave one product, $3'$-C-methylcytidine, with rather good activity in the suppression of neurovaccinial infections. The mechanism(s) by means of which these methylated nucleosides inhibit vaccinia virus have not as yet been studied. Nor, apparently, have they been tested against other viral systems.

The possible anti-viral activity of $O'$-alkylated nucleosides or nucleotides does not appear to have been investigated, notwithstanding that the $2'$- and $3'$-$O$-methyl adenosines, cytidines and uridines have been known for several years; while the existence of such residues in tRNA and rRNA is relatively widespread, and known to be due to the action of specific enzymes (see ref. [56] for review). Furthermore, ethionine-induced hepatic carcinoma in the rat is known to be accompanied by ethylation of tRNA, principally the sugar residues, i.e. the $2'$-OH [57, 58]. At least one report appears to confirm ethylation of

DNA under these conditions, the site of ethylation including probably the ring $N_7$ of guanosine [59].

Our interest in the $O'$-alkyl nucleosides stemmed initially from their use in the form of the 5'-pyrophosphates as potential substrates for polynucleotide phosphorylase, with a view to preparing the corresponding 2'-$O$-methyl polynucleotides [60, 61, 62] in order to obtain a better appreciation of the role of the 2'-OH in nucleic acid structure and function [63, 64].

The importance of this problem led us to develop a new method for the selective $O'$-alkylation of the sugar hydroxyls in cytosine nucleosides and 5'-nucleotides, the corresponding uracil glycosides being then obtained in almost quantitative yield by simple deamination procedures. The alkylation technique is based on dialkylsulphate treatment in alkaline medium, under controlled conditions, of unprotected cytosine glycosides, so that there is little or no alkylation of the ring $N_3$ nitrogen. The detailed procedure, which is described elsewhere [65], has made it possible to prepare the 2'-, 3'- and 5'-$O$-methyl and ethyl nucleosides, as well as the three di-$O'$- and the tri-$O'$- alkylated analogues. The 5'-$O$-alkylated derivatives have already proven useful in the development of a substrate for pancreatic ribonuclease which is fully resistant to phosphodiesterase I [66]. The potential anti-metabolic properties of this series of compounds will shortly be investigated.

Another procedure for the preparation of 2'- and 3'-$O$-methyl nucleosides has been recently reported by Robins and Naik [67, 68]. It is based on an extension of the observation of Aritomi and Kawasaki [69] that diazomethane treatment of $C$- and $O$-D-glucopyranosides in methanol, in the presence of catalytic amounts of $SnCl_2$, results in a considerable enhancement in the yield of $O$-methyl derivatives. In the presence of tin, the reaction also exhibits some degree of selectivity towards the sugar hydroxyls.

As will be shown below, 2'-$O$-alkylated polyribonucleotides are also of potential significance as viral and tumour anti-metabolites. It is therefore of interest that dialkylsulphate treatment of cytidine in alkaline medium [65], referred to above, is equally applicable to the 5'-phosphates, giving, in the case of cytidine-5'-phosphate, a mixture of the 2'- and 3'-$O$-methylcytidine-5'-phosphates, in high yield. These are then converted by standard procedures to the corresponding 5'-pyrophosphates, potential substrates of polynucleotide phosphorylase.

## $O'$-ALKYLATED ARABINOFURANOSYL NUCLEOSIDES

The application of the foregoing, and other techniques now under study, to the $O'$-methylation of arabinosyl cytosine and uracil nucleosides will undoubtedly provide a number of new and interesting derivatives of this important anti-viral agent. Furthermore, since, as pointed out above, ara-C is readily deaminated *in vivo* to the relatively inactive arabinofuranosyluracil, it is conceivable that

$O'$-alkylated ara-C may be less susceptible to enzymatic deamination, while at the same time retaining its activity as an antimetabolite.

To our knowledge only one attempt has hitherto been made to selectively methylate the sugar hydroxyls of an arabinosyl nucleoside, viz. 1-β-D-arabinofuranosyluracil [70]. The procedure used was based on the methylation, with methyl iodide according to Kuhn et al. [71], of the selectively tritylated derivatives of arabinouracil (spongouridine). The $2'$- and $3'$-$O$ methyl derivatives were obtained in crystalline form, but both were also methylated on the ring $N_3$ nitrogens by virtue of the methylation technique employed, and are consequently likely to be of only limited interest from a biological point of view. This is exemplified by the observation (see article by Prusoff elsewhere in ths volume) that $N_3$ methylation of 5-iododeoxyuridine leads to an almost total loss of anti-metabolic activity.

We have undertaken the preparation of the specific $O'$-methyl (and ethyl) derivatives of ara-C by dialkylsulphate treatment of the latter in alkaline medium, as described for cytidine [65]. This is followed by bisulphite deamination of the isolated products to give the corresponding uracil arabinosides without alkylation of the uracil ring.

Another, more limited, procedure involves conversion of a $3'$- or $5'$-$O$-alkyl cytidine to the corresponding arabinosyl nucleoside via the $2,2'$-anhydro derivative. For example, Mr. J. Giziewicz and Mr. J. Kusmierek have shown that treatment of $5'$-$O$-methylcytidine [65] with partially hydrolysed POCl$_3$ in boiling ethyl acetate [72] gave 2,2'-anhydro-$5'$-$O$-methylcytidine (isolated as the hydrochloride, m.p. 254-256°C decomp.). Hydrolysis of this derivative with 1 M triethylammonium carbonate at 50°C overnight [73] was followed by isolation, in 50% overall yield, of $5'$-$O$-methyl-ara-C (as the hydrochloride salt, melting point 151-154°C).

Preliminary trials on susceptibility to deamination of some of the foregoing derivatives have been carried out by Dr. M. Swierkowski, using a crude preparation of mouse kidney cytidine deaminase. Under conditions where cytidine and ara-C underwent total deamination, $2'$-$O$-methylcytidine, $3'$-$O$-methylcytidine and $5'$-$O$-methylcytidine proved completely resistant, while $5'$-$O$-arabinosylcytosine was deaminated to the extent of less than 10%. Of some interest was the fact that 6-methylcytidine also proved resistant to enzymatic deamination under these conditions. This study is now being extended to various $O'$-alkylated analogues of ara-C.

## ALKYLATED POLYNUCLEOTIDES

The synthesis, by enzymatic methods, of poly $2'$-$O$-methylcytidylic acid [61, 64], originally intended purely for physico-chemical studies on the role of the $2'$-OH in determining the differences between poly (rC) and poly (dC), held out some hope of the development of an interferon inducer more effective than poly (rI) . poly (rC). This expectation was based on the fact that poly (2'-$O$MeC) is, as expected theoretically, completely resistant to ribonuclease and is also highly resistant to phosphodiesterases.

However, extensive tests by Dr. Hilton Levy of the National Institutes of Health, Bethesda, and Dr. E. De Clercq (one of the speakers at the Symposium) have shown that poly (rI). poly (2'-$O$MeC), as well as poly (2'-$O$MeC) alone, exhibit negligible or no activity in test systems using poly (rI) . poly (rC) and poly (rC) alone as controls. From a therapeutic point of view, this result was disappointing, although Colby and Chamberlin [74] claim that susceptibility to mammalian ribonucleases of the poly (rC) component of poly (rI) . poly (rC) is not an important factor.

It is, on the other hand, of considerable interest from the point of view of the mechanism of interferon induction, that methylation of the 2'-hydroxyls in poly (rC) renders it inactive, and this problem is to be studied in more detail in collaboration with Dr. De Clercq. One factor which may, in part, account for the inactivity of poly (2'-$O$MeC) is the observation that the melting point, $T_m$, of its double-stranded complex with poly (rI) is 6-7° lower than that for poly (rI) . poly (rC). It is, however, difficult to see how this could conceivably account for the almost complete inactivation observed in two different laboratories. The situation is even more puzzling in the light of the reported observation that the complex of synthetic poly (1-vinyl)cytosine with poly (rI) exhibits activity comparable to that for poly (rI) . poly (rC) [75]. No data were presented by Pitha and Pitha [75] as to the nature of the complex formed, but the suggestion was advanced that activity is due to the known tendency of poly (1-vinyl)cytosine to form aggregates, which are more readily taken up by the cells. But how then to account for activity of a 1-vinylcytosine polymer?

This point is to be further investigated by an examination of the activity of poly (2'-$O$MeU) [62], the complex of which with poly (rA) has a higher melting point than that of the corresponding poly (rA) . poly (rU). Meanwhile, it has been reported by Dunlop *et al.* [76] that the 2'-$O$-methyl analogues of poly (rU), poly (rC) and poly (rA) are all inactive as messengers in *in vitro* systems, but that the activity of the poly (rU) analogue is stimulated by neomycin as in the case of polydeoxyribonucleotides. This finding may very well be of relevance to the observed lack of activity of poly (2'-$O$MeC) as an interferon inducer. It will also be of interest to determine the activity, in an interferon-inducing system, of the recently described poly 2'-deoxy-2'-chlorocytidylic acid [77].

A more unequivocal answer to the foregoing might be anticipated from another direction. Poly 5-methylcytidylic acid [78] forms a double-stranded complex with poly (rI), the melting point of which is 16°C *higher* than that for poly (rI) . poly (rC). At the time we prepared poly 5-methylcytidylic acid, the interferon-inducing ability of poly (rI) . poly (rC) had not been discovered. It is nonetheless surprising that others have not investigated the activity of poly

5-methylcytidylic acid, and we have consequently undertaken its preparation once again for this purpose. Simultaneously, we have been engaged in the preparation of poly 5-ethylcytidylic acid, which has been equally found to form a double-stranded complex with poly (rI), as presented by T. Kulikowski at this FEBS Meeting [79]. It is hoped, with the collaboration of Dr. De Clercq, to test this complex as well for interferon-inducing ability. In contrast to the 2'-O-methyl polynucleotides (see above), this polymer should be active as a messenger, since it has been shown by Mr. L. Novak in our laboratories that poly 5-ethyluridylic acid will readily code for phenylalanine in *in vitro* systems in the presence of a more elevated level of $Mg^{2+}$.

## ACKNOWLEDGEMENTS

We are indebted to Dr. C. Heidelberger and Dr. H. E. Kaufman for the tests of anti-viral activity; and to Dr. Hilton B. Levy and Dr. Erik De Clercq for the assay of interferon-inducing ability. The work reported herein profited from the support of the Polish Academy of Sciences (Project No. 09.3.1), The Wellcome Trust, the World Health Organization, and the Agricultural Research Service of the U.S. Department of Agriculture. The author is also indebted to the Medical Research Council of Canada for a Visiting Scientist grant during the preparation of this manuscript in the Département de Biochimie, Faculté de Médecine, Université Laval, Québec, Canada.

## REFERENCES

1. Prusoff, W. H. (1967). *Pharmac. Rev.* **19**, 209.
2. Goz, B. and Prusoff, W. H. (1970). *A. Rev. Pharmac.* **10**, 143.
3. Antiviral Substances (Symposium), (1970). *Ann. N.Y. Acad. Sci.* **173**, Article 1, pp. 1-844.
4. Roy-Burman, P. (1970). "Analogues of Nucleic Acid Components," Springer-Verlag, Heidelberg.
5. Borodkin, S., Johnson, S., Cocolas, H. and McKee, R. L. (1967). *J. med. Chem.* **10**, 248.
6. Ueda, T. and Kato, S. (1962). Patent (1961). 14,625; *Chem. Abstr.* **56**, 10170c.
7. Ueda, T. and Kato, S. (1962). Patent (1961) 13,826; *Chem. Abstr.* **56**, 10170f.
8. Gauri, K. and Rhode, B. (1969). *Klin. Wochenschr.* **47**, 375.
9. Cheng, C. C. (1969). *Progr. Med. Chem.* **6**, 67.
10. Cheng, C. C. and Roth, B. (1970). *Progr. Med. Chem.* **7**, 285.
11. Muraoka, M., Takada, A. and Ueda, T. (1962). *Keio J. Med.* **11**, 95.
12. Muraoka, M., Takada, A. and Ueda, T. (1970). *Chem. pharm. Bull. Tokyo* **16**, 261.
13. Muraoka, M., Seto, Y. and Ueda, T. (1970). *Chem. pharm. Bull. Tokyo* **16**, 269.
14. Piechowska, M. and Shugar, D. (1965). *Biochem. biophys. Res. Commun.* **20**, 708.

15. Brockman, R. W. and Anderson, E. P. (1963). *In* "Metabolic Inhibitors", (M. Rochstel and J. H. Quastel, eds), Academic Press, New York and London.
16. Slechta, L. and Hunter, J. H. (1970). *Ann. N.Y. Acad. Sci.* **173**, 708.
17. Sekely, L. and Prusoff, W. H. (1966). *Nature, Lond.* **211**, 1260.
18. Heidelberger, C. (1965). *Prog. Nucl. Acid Res.* **4**, 1.
19. Heidelberger, C. (1967). *A. Rev. Pharmac.* **7**, 101.
20. Barrett, H. W., Munavalli, S. N. and Newmark, P. (1965). *Biochim. biophys. Acta* **91**, 199.
21. Kaufman, H. E. (1965). *Prog. Med. Virol.* **7**, 116.
22. Hughes, W. F. (1969). *Am. J. Ophthalmol.* **67**, 313.
23. Kabat, S. and Visser, D. W. (1964). *Biochim. biophys. Acta* **80**, 680.
24. Nemes, M. M. and Hilleman, M. R. (1965). *Proc. Soc. exp. Biol. Med.* **119**, 515.
25. Kaufman, H. E. (1962). *Proc. Soc. exp. Biol. Med.* **109**, 251.
26. Kaufman, H. E. and Heidelberger, C. (1964). *Science, N.Y.* **145**, 585.
27. Walker, W. E., Waislren, B. A., Martins, R. R. and Batayias, G. E. (1971). "Antimicrobial Agents and Chemotherapy", p. 380.
28. Shugar, D., Swierkowski, M., Fikus, M. and Barszcz, D. (1967). 7th International Congress Biochem., Tokyo, Vol. I, Sym. 1, pp. 59-60.
29. Swierkowski, M. and Shugar, D. (1970). *J. molec. Biol.* **47**, 57.
30. Swierkowski, M. and Shugar, D. (1969). *Acta Biochim. Pol.* **16**, 263.
31. Swierkowski, M. and Shugar, D. (1969). *J. Med. Chem.* **12**, 533.
32. Pietrzykowska, I. and Shugar, D. (1966). *Biochem. biophys. Res. Commun.* **25**, 267.
33. Pietrzykowska, I. and Shugar, D. (1967). *Acta Biochim. Pol.* **14**, 165.
34. Gauri, K. K., Pflughaupt, K. W. and Müller, R. (1969). *Z. Naturf.* **24b**, 833.
35. Freese, E. (1959). *Proc. natn. Acad. Sci. U.S.A.* **45**, 622.
36. Künkel, H. A., Gauri, K. K. and Malorney, G. (1968). *Biophysik* **5**, 88.
37. Kaufman, H. E. (1971). Personal communication.
38. Gauri, K. K., Malorney, G. and Shiff, W. (1969). *Chemotherapy* **14**, 129.
39. Diwan, A. R., Robins, R. K. and Prusoff, W. H. (1969). *Experientia* **25**, 98.
40. Robins, R. K. (1971). Personal communication.
41. Winkley, M. W. and Robins, R. K. (1968). *J. org. Chem.* **33**, 2821.
42. Prystas, M. and Sorm, F. (1968). *Colln. Czech. Chem. Commun. Enol. Edr.* **34**, 2316.
43. Otter, A. B., Taube, A. and Fox, J. J. (1971). *J. Org. Chem.* **36**, 1251.
44. Pichat, L., Goldbillon, J. and Herbert, M. (1970). 7th Intern. Symp. Chem. of Natural Products (IUPAC), Riga (USSR), Abstract No. B39.
45. Ulbricht, T. L. V. (1959). *Tetrahedron* **6**, 225.
46. Kulikowski, T. and Shugar, D., (in preparation).
47. Woodman, R. J. (1966). *Nature, Lond.* **209**, 1362; *Cancer Res.* **28**, (1968) 2007.
48. Sidwell, R. W., Arnett, G. and Brockman, R. W. (1970). *Ann. N.Y. Acad. Sci.* **173**, (Art. 1) 592.
49. Kulikowski, T. and Shugar, D. (1971). *Acta Biochim. Pol.* **18**, 209.
50. Shen, T. Y. (1970). *Ange. Chem. Ausg. B.* **9**, 678.
51. Steuart, C. D. and Burke, P. J. (1971). *Nature (New Biology), Lond.*, **233**, 109.
52. Cohen, S. S. (1966). *Prog. Nucl. Acid. Res.* **5**, 1.

53. Sartorelli, A. C. and Creasy, W. A. (1968). *A. Rev. Pharmac.* **9**, 51.
54. Walton, E., Jenkins, S. R., Nutt, R. F. and Holly, P. W. (1969). *J. Med. Chem.* **12**, 306, and references cited therein.
55. Harper, P. J. and Hampton, A. (1970). *J. Org. Chem.* **35**, 1688.
56. Hall, R. H. (1970). "The Modified Nucleosides in Nucleic Acids", Columbia University Press, New York.
57. Farber, E. (1963). *Adv. Cancer Res.* **7**, 383.
58. Ortwerth, B. J. and Novelli, G. D. (1969). *Cancer Res.* **29**, 380.
59. Swann, P. F., Pegg, A. E., Hawks, A., Farber, E. and Magee, P. N. (1971). *Biochem. J.* **123**, 175.
60. Rottman, F. and Heinlein, K. (1968). *Biochemistry* **7**, 2634.
61. Janion, C., Żmudzka, B. and Shugar, D. (1970). *Acta Biochim. Pol.* **17**.
62. Żmudzka, B. and Shugar, D. (1970). *FEBS Letters* **8**, 52.
63. Bobst, A. M., Rottman, F. and Cerutti, P. A. (1969). *J. molec. Biol.* **46**, 221.
64. Żmudzka, B., Janion, C. and Shugar, D. (1969). *Biochem. biophys. Res. Commun.* **37**, 895.
65. Kuśmierek, J. and Shugar, D. (1971). *Acta Biochim. Pol.* **18**, 413.
66. Kole, R., Sierakowska, H. and Shugar, D. (1971). *Biochem. biophys. Res. Commun.* **44**, 1482.
67. Robins, M. J. and Naik, S. R. (1971). *Biochim. biophys. Acta* **246**, 341.
68. Robins, M. J. and Naik, S. R. (1971). *Biochemistry* **10**, 3591.
69. Aritomi, M. and Kawasaki, T. (1970). *Chem. pharm. Bull. Tokyo* **18**, 677.
70. Codington, J. P., Cushley, R. J. and Fox, J. J. (1970). *J. Org. Chem.* **35**, 3981.
71. Kuhn, R., Trischmann, H. and Low, I. (1955). *Angew. Chem.* **67**, 32.
72. Kanai, T., Kojima, T. and Maruyamo, O. (1970). *Chem. Pharm. Bull. Tokyo* **18**, 2569.
73. Kanai, T. and Ichino, M. (1971). *Tetrahedron Letters.* **1965**.
74. Colby, C. and Chamberlin, M. J. (1969). *Proc. natn. Acad. Sci. U.S.A.* **63**, 160.
75. Pitha, J. and Pitha, P. M. (1971). *Science, N.Y.* **172**, 1146.
76. Dunlop, B. E., Friderici, K. H. and Rottman, F. (1971). *Biochemistry* **10**, 2581.
77. Hobbs, J., Sternbach, H. and Eckstein, F. (1971). *FEBS Letters* **15**, 345.
78. Szer, W. and Shugar, D. (1966). *J. molec. Biol.* **17**, 174.
79. Kulikowski, T. and Shugar, D. (1971). Abs. Commun. 7th Meet. Fed. Europ. Biochem. Soc., p. 138.

## DISCUSSION

**P. Langen**: What is the basis of the anti-viral activity of EtUdR? Is it the labilization of DNA resulting from its incorporation, as shown by the lower melting point?

**D. Shugar**: No information is as yet available on the biochemical basis of the anti-viral activity of EtUdR. It seems most unlikely that the decrease in melting point of DNA, resulting from replacement of thymidine residues by EtUdR, could explain the anti-viral activity of the latter. It should be recalled that, in phage $T_3$ DNA where 70% of the thymidine residues have been replaced by EtUdR, the decrease in the $T_m$ value is only 4.5°C. What is required rather is a detailed investigation similar to that described in Dr. Prusoff's presentation on IUdR.

**J. Kaneti:** What is your opinion concerning the possibilities of relating calculated conformational transition energies to the experimentally observed values of the melting points for such polymers as poly (rU), poly (rT), poly (EtU), etc?

**D. Shugar:** I imagine you could better answer your own question than I could. However, I should mention that a small group of theoretical physicists at the Dept. of Biophysics of the University of Warsaw is at present working on this problem under the direction of Prof. W. Kolos.

**E. De. Clercq:** With regard to the substitution of the hydrogen on the 2'-OH (and 3'-OH) in uracil and cytosine nucleosides by methyl or propyl, is it possible to obtain dinucleoside monophosphates or trinucleoside monophosphates of these analogues? Furthermore, can these $O'$-methylated nucleosides be polymerized?

**D. Shugar:** The procedure we have developed, and the account of which is now in press (J. Kusmierek and D. Shugar, *Acta Biochim. Pol.* **18**, No. 4, 1971) is applicable only to the preparation of the 2'- (and 3'-) *O*-methyl and *O*-ethyl derivatives. No particular difficulties should be involved in the application of standard synthetic procedures to the preparation of di- and tri-nucleotides of these nucleoside analogues.

As regards the preparation of polynucleotides, this has already been achieved for some 2'-*O*-methyl analogues by the use of the 5'-pyrophosphates and polynucleotide phosphorylase. Attempts are currently under way in our laboratory by Dr. C. Janion to polymerize 2'-*O*-ethyl CDP and UDP, but hitherto without success. As regards polymers of the 3'-*O*-alkyl nucleosides, this would be possible only if enzymes of appropriate specificity were to be found, since chemical procedures would be very tedious.

**R. Roychoudhury:** Is it possible to replace the 3'-OH group in a deoxynucleoside by a 3'-*O*-methyl group, and would it be feasible to introduce this 3'-*O*-methylnucleoside as a terminal residue in an oligodeoxynucleotide?

**D. Shugar:** Technically, the preparation of 3'-*O*-methyldeoxynucleosides is considerably simpler than for the corresponding ribonucleosides, and we have already prepared several such analogues. There should be no difficulties involved in the chemical synthesis of oligodeoxyribonucleotides with such terminal groups; this should also be feasible by enzymatic procedures and this is also being attempted in our laboratories.

**W. H. Prusoff:** (a) Is EtUdR incorporated into mammalian DNA? (b) Is there any evidence for development of resistance to EtUdR? (c) Is there any evidence for toxicity of EtUdR in animal systems? (d) Has EtCdR been tested for anti-viral activity?

**D. Shugar:** (a) We have no information on this as yet, but Dr. Krystyna Swierkowska is at present studying this with the use of several cell lines. H. A. Künkel, K. K. Gauri and G. Malorney (*Biophysik* **5**, 88, 1968) have claimed an absence of any mutagenic effect of EtUdR on *Drosophila*, but it is difficult to ascribe any significance to this finding since no evidence was presented for EtUdR incorporation into DNA.

(b) No information is as yet available on this subject.

(c) Toxicity in animal systems has not yet been investigated, but we are synthesizing larger quantities of EtUdR for this purpose.

(d) The synthesis of EtCdR (5-ethyldeoxycytidine) is just nearing completion with the separation and crystallization of the *alpha* and *beta* anomers. We hope to have this available for testing within a few weeks and, in fact, we are hoping you can help us with these tests.

**W. H. Prusoff:** I should like to mention that EtUdR may prove of particular value in therapy of herpes encephalitis, since it will probably be considerably less toxic than IUdR. The concentrations of IUdR required for such therapy are quite toxic, but such toxicity has hitherto been regarded as acceptable in relation to the beneficial effects achieved.

## NOTE ADDED IN PROOF

Since submission of the foregoing manuscript, we have extended the procedure for $O'$-alkylation of cytidine [65] to arabinosylcytosine (ara-C), to obtain all possible $O'$-methyl analogues of ara-C. The three $O'$-monomethyl derivatives have been obtained in crystalline form and identified by various methods, including NMR spectra.

# Author Index

Numbers followed by an asterisk are those pages on which references are listed.

## A

Aaronson, S. A., 119 (17, 38), 120 (38, 41), 121 (17), 131*, 132*
Absher, M., 71 (1), 77 (113), 78 (1), 81*, 84*
Achong, B. G., 27 (2), 30*
Acs, G., 113 (27), 113*
Adams, R. Z. P., 100 (11), 101 (11), 113*
Adamson, R. H., 71 (2), 78 (2), 81*
Adesnik, M., 45 (1), 61*
Adye, J. C., 168 (24), 170 (24), 172 (24), 175*
Ahmed, M. S., 128 (33), 132*
Alarcon, R. A., 149 (8, 9), 161*
Aldrich, C., 87 (23), 98*
Allen, P. T., 119 (40), 132*
Allner, K., 47 (4), 61*
Aloni, Y., 35 (1), 37 (1), 43*
Altenburg, E., 2 (1), 6 (1), 10*
Amiel, J. L., 75 (91), 84*
Anderson, E. P., 194 (15), 204*
Anderson, W., 60 (10), 61*
Anebner, R. J., 94 (7), 97*
Aoki, T., 17 (1), 24*
Appell, L. H., 74 (58), 82*
Argaman, M., 177 (1), 190*
Aritomi, M., 200 (69), 205*
Armstrong, D., 27 (1), 30*
Armstrong, J. A., 77 (114, 115), 84*
Arnett, G., 198 (48), 204*
Asofsky, R., 76 (15), 81*
Asso, J., 50 (24), 62*
August, J. T., 60 (2), 61*
Austin, J. B., 94 (7), 97*
Aviv (Greenshpan), H., 45 (50, 51), 60 (50, 51), 62*
Axelrad, A. A., 2 (2), 3 (2), 10*
Axelrod, D., 37 (12), 44*

## B

Babcock, V. I., 74 (64), 83*
Bachrach, U., 149 (2, 5, 6, 10, 11, 12), 150 (16, 17, 18), 151 (12, 19, 20, 21, 22), 152 (20, 23, 24, 25), 153 (19, 25), 159 (5, 19), 161*
Bader, J. P., 100 (9), 104 (9), 113*
Bakhle, Y. S., 141 (41), 142 (1), 143 (40), 145*, 146*
Baldwin, R. L. 81*
Baltimore, D., 7 (3), 10*, 47 (5), 50 (23, 24), 61*, 62*, 115 (5), 119 (5), 120 (5), 131*, 138 (2), 145*
Banks, G. T., 66 (3, 4), 71 (3, 4), 81*
Banton, B. W., 139 (10), 145*
Barbanti-Brodano, G., 42 (2), 43*
Barmak, S., 78 (5), 81*
Baron, S., 67 (6, 11, 41), 73 (98, 130), 74 (16, 18, 58, 68, 69, 98, 130), 75 (68, 111), 77 (11, 42), 78 (90), 81*, 82*, 83*, 84*, 85*
Barr, Y. M., 27 (2), 30*
Barrett, H. W., 194 (20), 204*
Bart, R. S., 75 (7), 81*
Barton, B. W., 137 (3), 138 (3), 144*
Barszcz, D., 196 (28), 204*
Batayias, G. E., 195 (27), 204*
Bauer, D. J., 139 (48), 145*
Bauer, H., 99 (3), 113*
Bausek, G. H., 71 (8), 77 (8), 78 (9), 81*
Bayliss, N. L., 67 (88), 84*
Beard, D., 99 (2), 104 (17), 112 (2), 113*
Beard, J. W., 99 (1, 2), 104 (17), 112 (2, 26), 113 (26), 113*
Beaudreau, G. S., 99 (1, 4, 7), 100 (7), 104 (7), 113*, 119 (36), 132*
Becker, Y., 181 (32), 191*
Beckler, C. E., 74 (69), 83*
Béladi, I., 94 (1), 97*

209

Bellanti, J., 74 (69), 83*
Bello, L. J., 93 (16), 97*
Benneth, L. L., 170 (23), 172 (23), 173 (23), 175*
Ben-Porat, T., 142 (30), 146*, 187 (39, 40), 191*
Bentvelzen, P., 1 (4, 47), 2 (45, 64), 3 (5, 9, 18, 47, 52), 4 (4, 6, 9, 11, 47), 6 (10, 27), 7 (27), 7 (52), 8 (4, 7), 9 (9, 13), 10*, 11*, 12*
Berg, P., 81*
Berman, B. J., 47 (49), 62*
Bessell, C. J., 66 (10), 78 (10), 81*
Bessman, M. S., 126 (28), 132*
Bhuyan, B. K., 75 (135), 85*
Biczo, G., 165 (13), 175*
Bieber, A. L., 168 (20), 169 (20), 171 (20), 172 (20), 175*
Bierwolf, D., 15 (2), 24*
Billiau, A., 67 (6, 11), 75 (120), 77 (11), 78 (12), 81*, 85*
Birch, P. J., 66 (101), 84*
Bishop, J. M., 104 (18), 113*
Bittner, J. J., 1 (12), 11*
Black, F. L., 138 (16), 146*
Black, P. H., 35 (3), 43*, 94 (7), 97*
Blaskovic, D., 140 (43), 146*
Bobst, A. M., 200 (63), 205*
Bode, W., 45 (18), 61*
Bodo, G., 60 (20), 61*
Boedtker, H., 68 (43), 82*
Bogomolova, N. N., 67 (6), 81*
Bolle, A., 87 (5), 97*
Bolling, N. J., 66 (10), 78 (10), 81*
Bollum, F. J., 126 (29), 132*
Bolognesi, D. P., 99 (2, 3), 112 (2), 113*
Bonar, R. A., 99 (2), 112 (2), 112 (26), 113 (26), 113*
Borek, E., 45 (63), 63*
Borodkin, S., 193 (5), 203*
Bourali, C., 75 (61), 82*, 83*
Bowen, J. M., 119 (40), 132*
Bowman, C. M., 45 (3), 61*, 63*, 64 (66)
Boy de la Tour, E., 87 (5), 97*
Boyer, P. D., 166 (17), 175*
Boyse, E. A., 7 (58), 8 (58), 12*, 16 (15), 17 (1), 18 (4), 21 (26), 25*, 27 (15), 31*

Brachet, J., 181 (29), 191*
Bradish, C. J., 47 (4), 61*
Brandt, J., 76 (87), 78 (87), 84*
Braun, W., 76 (128), 85*
Brawerman, G., 59 (36), 62*
Breeden, C. J., 139 (4), 145*
Brennan, M. J., 128 (35), 132*
Brenner, S., 87 (12), 97*
Breyere, E. J., 21 (3), 24*
Brihaye, J., 140 (44), 146*
Brockman, R. W., 194 (15), 198 (48), 204*
Brockman, W. W., 180 (21), 190*
Brommer, E. J. P., 9 (13), 11*
Brown, D. C., 137 (5), 145*
Brox, L. W., 166 (16), 175*
Bruyn, J., de, 189 (42), 191*
Buck, K. W., 66 (3, 4, 13), 71 (3, 4), 81*
Buckler, C. E., 67 (6, 11, 41), 77 (11, 42), 81*, 82*
Bundeally, A. E., 78 (99), 84*
Burdette, W. J., 3 (66), 9 (14), 11*, 12*
Burke, D. C., 67 (55, 56, 89), 82*, 84*
Burke, P. J., 199 (51), 204*
Burness, A. T. H., 45 (6), 61*
Burny, A., 7 (57), 12*, 104 (19), 108 (19), 113*, 115 (6, 7), 116 (6), 117 (6), 118 (6), 119 (6, 7, 25), 120 (20), 121 (7, 20), 123 (20), 124 (20, 25), 125 (25), 126 (27), 127 (27), 128 (6), 131*, 132*
Burrows, J. H., 128 (35), 132*
Burt, W. L., 78 (97), 84*
Bush, H., 110 (21), 111 (21), 112 (21), 113*
Butel, J. S., 33 (4), 43*
Butterworth, B. E., 47 (7), 50 (7), 61*

C

Calabresi, P., 138 (28), 139 (6, 7), 145*, 146*
Calafat, J., 2 (64), 3 (9), 4 (9), 6 (27), 7 (27), 9 (9), 11*, 12*
Calvin, M., 181 (25), 191*
Came, P. E., 74 (125), 75 (14), 81*, 85*

## AUTHOR INDEX

Camerman, N., 143 (9), 145*
Campbell, A., 87 (2), 97*
Canellakis, E. S., 59 (38), 62*
Cannellos, G., 74 (69), 83*
Cantagalli, P., 77 (39, 40), 82*
Cantor, H., 76 (15), 81*
Carbone, P., 74 (69), 83*
Cardoso, S. S., 139 (6), 145*
Carnegie, J. W., 99 (4), 113*
Carter, W. A., 46 (8, 9, 37), 55 (9), 61*, 62*, 180 (21), 190*
Catalano, L. W., Jr., 74 (16), 81*
Cavalieri, L. F., 124 (26), 132*
Cerutti, P. A., 200 (63), 205*
Chagoya, V., 142 (24), 146*
Chain, E. B., 66 (3, 4, 13), 71 (3, 4), 81*
Chalvet, O., 164 (4), 175*
Chamberlin, M., 81*
Chamberlin, M. J., 68 (19), 69 (19), 70 (19), 77 (19), 81*, 202 (74), 205*
Chan, S. P., 76 (117), 84*
Chang, P. K., 141 (42), 146*, 147*
Chanock, R. M., 74 (69), 83*
Charney, J., 128 (35), 132*
Chase, M., 152 (26), 159 (26), 161*
Chen, B., 60 (10), 61*
Cheng, C. C., 193 (9, 10), 203*
Chevalley, E., 87 (5), 97*
Chirigos, M. A., 76 (117), 77 (109), 84*
Choay, J., 75 (91), 84*
Chopra, H. C., 128 (31, 32, 33), 132*
Chowchuvech, E., 74 (18), 81*
Chu, M. Y., 139 (6), 145*
Chubb, R. C., 8 (53), 12*
Clark, W. R., 75 (80), 83*
Clarke, J. K., 112 (24), 113*
Cleaver, J. E., 184 (38), 191*
Clifford, P., 27 (6, 15), 28 (18), 30*, 31*
Cocolas, H., 193 (5), 203*
Codington, J. P., 201 (70), 205*
Cohen, N., 47 (30), 62*
Cohen, S. S., 199 (52), 204*
Colby, C., 67 (20), 68 (19), 69 (19), 70 (19, 21), 77 (19), 81*, 202 (74), 205*
Cole, R. K., 4 (15), 11*, 115 (2), 131*
Colter, S. S., 93 (9), 97*

Conchie, A. F., 139 (10), 145*
Constantino, L., 149 (3), 161*
Conway, T. W., 45 (53), 62*
Coppey, J., 75 (62), 83*
Coward, J. E., 120 (15), 131*
Cramer, J. W., 137 (11), 145*
Crawford, L. V., 45 (43), 62*
Creasy, W. A., 199 (53), 205*
Crescenzi, V., 149 (3), 161*
Cricchio, 182 (35), 191*
Crippa, M., 181 (28, 301), 191*
Crittenden, L. B., 3 (16, 17), 11*
Crombrugghe, B., de, 60 (10), 61*
Cushley, R. J., 201 (70), 205*
Cysyk, R., 144 (8), 145 (8a, 55a), 145*, 147*

### D

Daams, J. H., 3 (9), 4 (9, 11), 6 (27), 7 (27), 8 (7), 9 (9), 10*, 11*
Dagle, G. E., 83*
Dahlberg, J. E., 45 (3), 61*
Dale, B., 45 (18), 61*
Dales, S., 177 (5), 190*
Daniel, V., 45 (11), 61*
Darbyshire, J. E., 66 (4), 71 (4), 81*
Darnell, J. E., 59 (12), 61*
Das, M. R., 7 (57), 12*, 115 (6), 116 (6), 117 (6), 118 (6), 119 (6, 25), 120 (20), 121 (20), 123 (20), 124 (20, 25), 125 (25), 128 (6), 131*
Datta, S. K., 75 (120), 85*
Daudel, R., 164 (4), 175*
Davis, R. E., 163 (3), 175*
Dawson, C. R., 78 (97), 84*
De Clercq, E., 68 (22, 31, 35), 69 (30, 32, 35, 36), 70 (31, 32, 36, 37), 71 (27, 28, 37), 73 (23, 24, 34), 74 (23, 34), 76 (29), 77 (24, 26, 30, 31, 32, 33, 35, 36), 78 (25, 27), 81*, 82*
Deeney, A. O. C., 99 (4), 113*
De Maeyer, E., 66 (38), 82*
De Maeyer-Guignard, J., 66 (38), 78 (72), 82*, 83*
Demissie, A., 28 (18), 31*
Denhardt, D. T., 122 (23), 123 (23), 131*

Denhardt, G. T., 87 (5), 97*
Deringer, M. K., 3 (30), 11*
Dermott, E., 112 (24), 113*
De Santis Savino, M., 149 (3), 161*
De Somer, P., 71 (27, 28), 75 (120), 76 (29), 78 (12, 27), 81*, 85*
De Vassal, F., 75 (91), 84*
De Vita, V., 74 (69), 83*
Dianzani, F., 67 (11, 41), 77 (11, 39, 40, 42), 81*, 82*
Diggelmann, H., 181 (24), 190*
Dingman, C. W., 103 (15), 110 (15), 113*
Diwan, A., 135 (12), 142 (13, 14), 145*
Diwan, A. R., 197 (39), 204*
Dmochowski, L., 119 (40), 132*
Dobos, P., 47 (13), 48 (13), 50 (13), 54 (13), 61*
Don, S., 150 (18), 151 (19, 20), 153 (19), 159 (19), 161*
Doty, P., 68 (43), 82*, 92 (21), 98*
Douthart, R. J., 67 (75), 83*
Dubbs, D. R., 33 (9), 34 (10), 43*
Dube, S. K., 45 (14, 58), 61*, 63*
Dubois-Dalcq, M., 189 (41), 191*
Duesberg, P. H., 66 (73), 67 (20), 81*, 83*
Dulbecco, R., 35 (13), 37 (13), 44*
Dunlop, B. E., 202 (76), 205*
Dunn, T. B., 3 (30), 11*
Dux, A., 3 (18, 19, 20), 11*

E

Earhart, C. F., 45 (46), 62*
Eckstein, F., 68 (31, 35), 69 (30, 35), 70 (31), 77 (30, 31, 35), 82*, 202 (77), 205*
Eccleston, E., 78 (83), 83*
Edgar, R. S., 87 (3, 5), 97*
Edmonds, M., 59 (15), 61*
Eilon, G., 149 (3), 150 (16), 151 (20, 21, 22), 152 (20), 161*
Eker, R., 1 (21), 11*
Elia, V., 149 (3), 161*
Elion, G. B., 165 (10), 166 (18), 168 (19), 169 (19), 170 (18, 19), 171 (18, 19), 172 (19), 175*

Elis, J., 140 (37), 146*
Ellis, L. F., 66 (74), 83*
Emmelot, P., 2 (45), 3 (18, 52), 4 (6), 6 (27), 7 (27, 52), 10*, 11*, 12*
Engelberg, H., 179 (17), 190*
Engelhardt, D. L., 59 (42), 62*
Ensinger, M. J., 87 (4), 97*
Epstein, M. A., 27 (2), 30*
Epstein, R. H., 87 (5), 97*
Epstein, S. S., 4 (22), 11*
Erickson, E. L., 100 (8), 113*
Ernberg, I., 28 (3), 29 (4, 5), 30*
Evans, A. S., 27 (14), 31*
Eyerman, E., 78 (131), 85*

F

Fabro, S., 71 (2), 78 (2), 81*
Falcoff, E., 66 (45, 46), 75 (62), 77 (44), 82*, 83*
Falcoff, R., 66 (45, 46), 75 (62), 82*, 83*
Falke, D., 140 (15), 146*
Fantes, K. H., 51 (16), 61*, 66 (10), 78 (10), 81*
Farber, E., 199 (57), 200 (59), 205*
Fekete, E., 1 (56), 4 (23), 11*, 12*
Fenje, P., 74 (47, 48), 82*
Fenner, F., 87 (6), 97*
Fey, F., 15 (2), 24*
Ficq, A., 181 (29), 191*
Field, A. K., 66 (50, 76, 94, 118), 67 (52), 68 (49, 51, 78, 119), 73 (93), 74 (93), 75 (93), 77 (77), 82*, 83*, 84*, 85*
Fikus, M., 196 (28), 204*
Finch, S. C., 139 (6), 145*
Finter, N. B., 46 (17, 57), 61*, 63*, 174 (53), 77 (53), 82*
Fischer, D. S., 138 (16), 146*
Fischer, H., 35 (5), 43*
Fleisher, M., 78 (134), 85*
Foft, J. W., 45 (22, 65), 61*, 63*
Follett, E. A. C., 177 (6), 179 (13), 181 (13, 33), 182 (13), 190*, 191*
Fontaine-Brouty-Boyé, D., 75 (62), 83*
Forbes, M., 74 (86), 78 (87), 83*, 84*
Force, E. E., 138 (17), 146*

# AUTHOR INDEX

Fox, J. J., 197 (43), 201 (70), 204*, 205*
Fox, T. O., 38 (6), 43*
Frearson, P. M., 33 (9), 43*
Freeman, A. E., 94 (7), 97*
Freese, E., 196 (35), 204*
Fresco, J. R., 68 (43), 82*
Friderici, K. H., 202 (76), 205*
Friedman, A., 76 (129), 85*
Friedman, R. M., 45 (57), 46 (57), 63*
Friedman-Kien, A. E., 70 (124), 74 (59), 75 (54), 77 (124), 82*, 83*, 85*
Friesen, J. D., 45 (18), 61*
Fujinaga, K., 94 (8), 96 (8), 97*, 104 (17), 113*, 119 (39), 132*
Fujita, T., 164 (9), 175*
Fukami, H., 149 (13, 14, 15), 161*
Fukui, K., 171 (25), 174 (27), 175*
Furth, J., 4 (15), 11*, 115 (2), 131*

## G

Gagnoni, S., 67 (41), 77 (39, 40), 82*
Galin, M. A., 74 (18), 81*
Gallerani, R., 60 (35), 62*
Gallo, R. C., 119 (40), 132*, 180 (19), 190*
Gandhi, S. S., 67 (55, 56), 82*
Gauri, K., 193 (8), 203*
Gauri, K. K., 140 (18, 19, 20), 146*, 196 (34, 36), 197 (38), 204*
Gavrilova, L. P., 101 (14), 102 (14), 104 (14), 108 (14), 109 (14), 113*
Gay, F., 112 (24), 113*
Gazdar, A. F., 75 (126), 85*
Geering, G., 18 (4, 5), 24*, 25*, 27 (15), 31*
Gelboin, H. V., 75 (57), 82*
Gergely, L., 28 (3, 12), 29 (4, 5), 30*
Gerone, P. J., 74 (58), 82*
Gharpure, M., 87 (27), 98*
Gifford, G. E., 67 (60), 83*
Giglio, E., 149 (3), 161*
Gilden, R. V., 18 (9), 25*, 119 (37), 132*
Gillespie, D., 122 (24), 131*
Ginsberg, H. S., 87 (4), 93 (16), 97*

Girard, M., 35 (7), 43*
Gober, L. L., 74 (59), 83*
Gold, E., 142 (47), 146*
Goldbillon, J., 197 (44), 198 (44), 204*
Goldblum, N., 150 (17), 161*, 177 (1), 190*
Goldblum, T., 150 (17), 161*
Goldstein, G., 28 (12), 30*
Golgher, R. R., 47 (49), 62*
Golovinsky, E., 174 (29), 175*
Goodman, N. C., 126 (30), 127 (30), 132*
Goorha, R. M., 67 (60), 83*
Gosser, L. B., 78 (112), 84*
Gots, J. S., 168 (24), 170 (24), 172 (24), 175*
Gottesman, M., 60 (10, 48), 61*, 62*
Goulian, M., 122 (22), 131*
Gowdy, C. N., 135 (12), 145*
Goz, B., 137 (23), 140 (23), 142 (13, 23), 143 (21, 22), 145*, 146*, 193 (2), 198 (2), 203*
Graffi, A., 15 (2), 24*
Grant, R. C., 69 (36), 70 (36), 77 (36), 82*
Green, M., 38 (17), 44* 87 (10), 91 (19), 92 (18), 94 (8), 96 (8), 97*, 98*, 104 (17), 113*, 119 (39), 132*, 142 (24), 146*, 179 (18), 180 (20), 190*
Gresland, L., 67 (70), 83*
Gresser, I., 75 (61, 62), 83*
Grimley, P. M., 177 (3, 4, 7, 8), 179 (16), 180 (16), 181 (16), 190*
Grimsson, H., 120 (10), 131*
Groner, Y., 45 (50, 51), 60 (50, 51), 62*
Gross, L., 3 (26), 4 (24, 25, 26), 11*
Grunwald, E., 164 (7), 175*
Guggenheim, M. A., 74 (69), 83*
Gugten, A. v.d., 4 (11), 11*
Gurgo, C., 119 (39), 132*, 179 (18), 190*

## H

Hackett, A. J., 181 (25), 191*
Hageman, P., 3 (9), 4 (9), 6 (27), 7 (27), 9 (9), 11*

Hageman, P. C., 2 (64), 12*
Hahn, E. C., 38 (15), 44*
Hairstone, M. A., 3 (28), 11*
Hakura, K., 17 (14), 25*
Hall, B. D., 117 (9), 131*
Hall, R. H., 199 (56), 205*
Hall, T. C., 139 (4), 145*
Hallum, J. V., 77 (132, 133), 85*
Hamilton, L. D., 74 (64), 78 (134), 83*, 85*
Hamilton, M. G., 47 (29), 62*
Hämmerling, U., 17 (1), 24*
Hammett, L. P., 164 (8), 175*
Hampton, A., 199 (55), 205*
Hanafusa, H., 181 (26), 191*
Handler, P., 166 (17), 175*
Hanna, C., 137 (25), 146*
Hanna, L., 74 (65), 83*
Hansch, C., 164 (9), 175*
Hansen, K., 170 (21), 172 (21), 173 (21), 175*
Haran-Ghera, N., 15 (6), 25*
Hardy, W. D., 18 (5), 24*
Harel, L., 66 (66), 83*
Harper, P. J., 199 (55), 205*
Harter, D. H., 115 (7), 119 (7), 120 (15), 121 (7, 18), 131*
Hartmann, G., 172 (26), 173 (26), 175*
Harven, E., de, 17 (1), 18 (5), 24*, 25*, 27 (15), 31*
Haselkorn, R., 68 (43), 82*
Hatanaka, M., 119 (37), 132*
Hausen, H., zur, 27 (6), 30*
Havell, E. A., 74 (59), 83*
Hawks, A., 200 (59), 205*
Hayat, M., 75 (91), 84*
Haynes, G. R., 142 (32), 146*
Heberline, R. L., 74 (67), 83*
Heidelberger, C., 194 (18, 19), 195 (26), 204*
Heine, U., 2 (42), 11*
Heinlein, K., 200 (60), 205*
Heller, E., 177 (1), 190*
Henderson, J. F., 166 (16), 175*
Henle, G., 27 (1, 6, 8), 28 (9, 10), 29 (7), 30*
Henle, W., 27 (1, 6, 8), 28 (9, 10), 29 (7), 30*
Henry, P. H., 94 (7), 97*
Herbert, M., 197 (44), 198 (44), 204*

Herman, R., 74 (68), 75 (68), 83*
Herrmann, E. C., Jr., 137 (27), 138 (26), 142 (26), 146*
Hershey, A. D., 152 (26), 159 (26), 161*
Heston, W. E., 1 (29), 3 (30), 11*
Hilfenhaus, J., 60 (20), 61*
Hill, D. A., 74 (58, 69), 82*, 83*
Hill, D. L., 170 (23), 172 (23), 173 (23), 175*
Hilleman, M. R., 66 (50, 76, 94, 118), 67 (52), 68 (49, 51, 78, 119), 73 (93), 74 (93), 75 (80, 81, 93), 77 (77), 82*, 83*, 84*, 85*, 195 (24), 204*
Hillova, J., 67 (70), 83*
Himmelweit, F., 66 (3, 4, 13), 71 (3, 4), 81*
Hiraki, K., 15 (10), 25*
Hirt, B., 40 (8), 43 (8), 43*
Hitchings, G. H., 166 (18), 170 (18), 171 (18), 175*
Ho, M., 77 (114, 115), 84*, 94 (11), 97*
Hobbs, J., 202 (77), 205*
Holland, I. B., 45 (55), 63*
Holland, J. J., 50 (19), 61*
Hollings, M., 66 (3), 71 (3), 81*
Holly, P. W., 199 (54), 205*
Homan, E., 74 (69), 83*
Horak, I., 60 (20), 61*
Horne, R. W., 87 (12), 97*
Hsu, W.-T., 45 (21, 22, 34, 65), 61*, 62*, 63*
Huang, A. S., 47 (5), 61*
Hudson, R. F., 163 (1), 175*
Huebner, R. J., 3 (31), 6 (31), 7 (31), 11*, 18 (7, 9), 20 (8), 25*, 35 (3), 43*, 75 (92, 108, 111), 84*, 94 (17), 98*, 119 (37), 132*, 138 (28), 146*
Hughes, W. F., 195 (22), 204*
Hunt, J. M., 59 (42), 62*
Hunter, J. H., 194 (16), 204*
Huppert, J., 67 (70), 83*

I

Ichino, M., 201 (73), 205*
Ikemura, T., 45 (3), 61*

Imsande, J., 166 (17), 175*
Irino, S., 15 (10), 25*
Ishibashi, M., 87 (13), 97*
Ishizaki, R., 3 (66), 12*
Ito, M., 87 (14), 98*
Iwasa, J., 164 (9), 175*

**J**

Jackson, J., 104 (18), 113*
Jacobson, M. F., 50 (23, 24), 61*, 62*
Jaffe, H. H., 164 (5), 175*
Jahiel, R. I., 74 (71), 83*
Janion, C., 200 (61, 64), 201 (61, 64) 205*
Jasmin, C., 75 (91), 84*
Jawetz, E., 74 (65), 83*
Jenkins, S. R., 199 (54), 205*
Jensen, E. M., 128 (32, 33), 132*
Jensen, F. C., 37 (11), 43*
Jensen, L. H., 165 (14), 175*
Johnson, S., 193 (5), 203*
Johnston, E. N. M., 139 (35), 146*
Joklik, W. K., 46 (25), 47 (25), 56 (25), 60 (25), 62*
Jones, D., 78 (83), 83*
Joss, U. R., 181 (25), 191*
Juel-Jensen, B. E., 138 (34), 139 (29), 146*
Jullien, P., 78 (72), 83*
Jungwirth, C., 60 (20), 61*

**K**

Kabat, S., 195 (23), 204*
Kacian, D. L., 126 (27), 127 (27), 132*
Kalf, G. F., 101 (12), 113*
Kalter, S. S., 74 (67), 83*
Kamen, R., 60 (26), 62*
Kan, J., 45 (27), 62*
Kanai, T., 201 (72, 73), 205*
Kaneti, J., 174 (29), 175*
Kano-Sueoka, T., 45 (59), 63*
Kapikian, A. Z., 74 (69), 83*
Kaplan, A. S., 142 (30), 146*, 187 (39, 40), 191*
Kaplan, H. S., 2 (32), 4 (32, 36), 11*, 15 (11), 25*
Kates, J., 59 (28), 62*
Kato, S., 193 (6, 7), 195 (7), 203*
Katz, E., 150 (17), 161*, 177 (3, 4, 9, 10), 190*
Kaufman, H. E., 138 (31), 146*, 195 (21, 25, 26), 197 (37), 204*
Kawasaki, H., 149 (13, 14, 15), 161*
Kawasaki, T., 200 (69), 205*
Ke, Y. H., 77 (114), 84*
Keir, H. M., 142 (47), 146*
Kellenberger, E., 87 (5), 97*
Kelley, W. N., 166 (16), 175*
Kelloff, G. J., 18 (9), 25*
Kerr, I. M., 45 (43), 46 (31, 32), 47 (13, 29, 30, 32, 33), 48 (13), 50 (13), 51 (33), 53 (32, 33), 54 (13, 32), 61*, 62*
Keydar, J., 7 (57), 12*, 115 (6), 116 (6), 117 (6), 118 (6), 119 (6, 25), 120 (20), 121 (20), 123 (20), 124 (20, 25), 125 (25), 128 (6), 131*
Kidwai, J. R., 37 (16), 38 (16), 44*
Kiehn, E. D., 50 (19), 61*
Kier, L. B., 174 (28), 175*
Kimball, P. C., 66 (73), 83*
Kimes, B. W., 149 (7), 161*
Kimes, R., 87 (6), 97*
Kirschstein, R. L., 74 (69), 83*
Kit, S., 33 (9), 34 (10), 43*
Klein, G., 27 (6, 11, 13, 17), 28 (3, 10, 12, 18), 29 (4, 5, 20), 30*, 31*
Kleinschmidt, W. J., 66 (74), 67 (75), 83*
Klem, E. B., 45 (34), 62*
Klemperer, H. G., 142 (32), 146*
Klenow, H., 170 (21), 172 (21), 173 (21), 175*
Kligerman, M. M., 139 (6), 145*
Klopman, G., 163 (1, 2, 3), 164 (2), 175*
Kohler, K., 94 (11), 97*
Kohn, G., 30 (21), 31*
Kojima, T., 201 (72), 205*
Kole, R., 200 (66), 205*
Kondo, M., 60 (35), 62*
Konisky, J., 45 (3), 61*
Kopf, A. W., 75 (7), 81*
Koprowski, H., 37 (11), 42 (2), 43*
Kornberg, A., 122 (21, 22), 131*
Kornberg, A. J., 126 (28), 132*
Korner, A., 47 (44), 62*
Korteweg, R., 2 (33), 11*

Kourilsky, F. M., 27 (17), 31*
Kramarsky, B., 128 (35), 132*
Kraut, J., 165 (14), 175*
Krawciw, T., 70 (124), 77 (124), 85*
Krenitsky, T. A., 166 (18), 168 (19), 169 (19), 170 (18, 19), 171 (18, 19), 172 (19), 175*
Kubinski, H., 143 (53), 147*
Kuhn, R., 201 (71), 205*
Kulikowski, T., 198 (46, 49), 203 (79), 204*, 205*
Künkel, H. A., 196 (36), 204*
Küntzel, H., 101 (13), 102 (13), 113*
Kurimura, T., 34 (10), 43*
Kusmierek, J., 200 (65), 201 (65), 205*

## L

Ladik, J., 165 (13), 175*
Lampson, G. P., 66 (50, 76, 94, 118), 67 (52), 68 (49, 51, 78, 119), 73 (93), 74 (93), 77 (77), 82*, 83*, 84*, 85*
Lamy, M., 182 (36), 191*
Lancini, G., 182 (34, 35), 191*
Lane, W. T., 18 (9), 25*, 138 (28), 146*
Lange, J., 18 (24), 25*
Langlois, A. J., 99 (2), 112 (2), 112 (26), 113 (26), 113*
Lardy, H., 166 (17), 175*
Larke, R. P. B., 75 (127), 85*
Larson, V. M., 75 (80, 81), 83*
Lasfargues, E. Y., 128 (35), 132*
Last, F. T., 66 (3), 71 (3), 81*
Laursen, A. C., 66 (10), 78 (10), 81*
Law, L. W., 3 (35), 4 (35), 11* 75 (85), 83*
Lawrence, W., Jr., 76 (104), 84*
Ledinko, N., 142 (33), 146*
Lee, S. Y., 59 (36), 62*
Lee-Huang, S., 124 (26), 132*
Leffler, J. E., 164 (7), 175*
Lehman, I. R., 126 (28), 132*
Leibach, T., 172 (26), 173 (26), 175*
Leibovici, J., 149 (12), 151 (12), 152 (25), 153 (25), 161*
Lemke, P. A., 66 (82), 83*
Leonard, B. J., 78 (83), 83*

Levin, M. J., 59 (47), 60 (47), 62*
Levine, A. J., 38 (6), 43*, 93 (16), 97*
Levine, L., 78 (99), 84*
Levinson, W., 104 (18), 113*
Levinthal, C., 45 (1), 61*
Levy, H., 76 (104), 84*
Levy, H. B., 46 (8, 9, 37), 55 (9, 39), 61*, 62*, 67 (6), 74 (69), 75 (57, 85, 126), 76 (15, 84), 77 (42), 78 (90), 81*, 82*, 83*, 84*, 85*
Levy, J. P., 27 (17), 31*, 83*
Lewis, A. M., Jr., 94 (17), 98*
Lieberman, M., 4 (36), 11* 15 (11), 25*
Lielausis, A., 87 (5), 97*
Lielausis, I., 87 (3), 97*
Lilly, F., 3 (37, 38, 39), 11*, 15 (12), 25*
Lim, L., 59 (38), 62*
Lin, F. H., 119 (16), 121 (16), 131*
Lindberg, B., 170 (21, 22), 172 (21, 22), 173 (21, 22), 175*
Lindh, H. F., 74 (86), 83*
Lindsay, H. L., 74 (86), 76 (87), 78 (87), 83*, 84*
Lindsay, J. G., 100 (11), 101 (11), 113*
Lipman, M. B., 142 (36), 145*
Liquori, A. M., 149 (3), 161*
Litt, M., 68 (43), 82*
Littauer, U. Z., 45 (11), 61*
Lockart, R. Z., Jr., 67 (88), 78 (112), 84*
Lodish, H. F., 112 (23), 113*
Loebenstein, G., 150 (16), 161*
Loening, U., 110 (22), 110 (22), 113*
Loening, U. E., 103 (16), 109 (16), 110 (16), 111 (16), 113*
Lomniczi, B., 67 (89), 84*
Louis, J. B., 163 (3), 175*
Low, I., 201 (71), 205*
Luzzati, D., 181 (27), 191*
Lwoff, A., 3 (40), 4 (40, 41), 11*

## M

Maca, R. A., 2 (42), 11*
McAllister, R. M., 94 (17), 98*

# AUTHOR INDEX

McAuslan, B. R., 55 (39), 62*, 179 (11), 190*
McCallum, D. I., 139 (35), 146*
MacCallum, F. O., 138 (34), 139 (29), 146*
McCollum, R. W., 27 (14), 31*, 138 (28), 139 (7), 145*, 146*
McCrea, J. F., 143 (36, 40), 146*
McDonald, 87 (27), 98*
MacHattie, L. A., 87 (10), 97*
McIntosh, K., 87 (20), 98*
McKee, R. L., 193 (5), 203*
Mackenzie, A. M. R., 139 (29), 146*
McKillop, T. F. W., 164 (4), 175*
MacPherson, I., 94 (17), 98*
MacPherson, I. A., 9 (43), 11*
Magee, P. N., 200 (59), 205*
Maizel, J. V., Jr., 50 (52), 62*, 87 (15), 98*
Malorney, G., 196 (36), 197 (38), 204*
Malorny, G., 140 (19), 146*
Manaker, R. A., 2 (42), 11*
Mantyjarvi, R., 93 (16), 98*
Maral, R., 182 (37), 191*
Marbaix, G., 104 (19), 108 (19), 113*
Marcker, K. A., 47 (56), 63*
Marcus, P. I., 46 (40, 41), 59 (32), 62*
Margolis, S. A., 78 (90), 84*
Marks, J. E., 66 (3), 71 (3), 81*
Marmur, J., 92 (21), 98*
Marquardt, H., 78 (134), 85*
Martin, E. M., 45 (43), 46 (31), 47 (13, 29, 33), 48 (13), 50 (13), 51 (33), 53 (33), 54 (13), 61*, 62*
Martin, M. A., 37 (12), 44*
Martins, R. R., 195 (27), 204*
Marty, L., 35 (7), 43*
Maruyama, K., 17 (14), 25*
Maruyamo, O., 201 (72), 205*
Mason, M. M., 128 (31), 132*
Mathe, G., 75 (91), 84*
Mathews, M. B., 47 (44, 56), 62*, 63*
Matsukura, M., 3 (50), 12*
Mayberry, B. R., 74 (86), 83*
Mayyasi, S. A., 128 (33), 132*
Mécs, I., 78 (110), 84*
Meier, H., 18 (9), 25*, 75 (92), 84*
Melnick, J. L., 2 (65), 12*, 33 (9), 43*

Mendecki, J., 59 (36), 62*
Merigan, T. C., 46 (25), 47 (25), 56 (25), 60 (25), 62*, 68 (22, 31, 35), 69 (30, 32, 35, 36), 70 (31, 32, 36, 37), 71 (8, 37), 73 (23, 24, 34), 74 (23, 34, 65, 105, 106), 77 (8, 24, 26, 30, 31, 32, 33, 35, 36), 78 (9, 25), 81*, 82*, 83*, 84*
Meyers, D. D., 18 (9), 25*
Micheel, B., 16 (20), 22 (13, 20), 25*
Miller, R. L., 168 (20), 169 (20), 171 (20), 172 (20), 175*
Miller, Z., 60 (45), 62*
Mills, J., 74 (69), 83*
Mims, S. J., 177 (4), 190*
Mitchell, W. M., 2 (44), 11*
Mizuno, N. S., 191*
Mizutani, S., 7 (63), 12*, 115 (4), 119 (4), 131*, 138 (54), 147*
Montagnier, L., 2 (45), 12*, 66 (38, 66), 82*, 83*
Montgomery, J. A., 140 (48), 146*
Moore, D., 115 (8), 119 (8), 129 (8), 131*
Moore, D. H., 1 (49), 2 (49), 3 (28, 46), 7 (49), 11*, 12*, 75 (14), 81*, 128 (35), 132*
Morgan, C., 120 (15), 131*
Morino, T., 149 (13, 14, 15), 161*
Morris, D. R., 149 (7), 161*
Moses, H. L., 2 (44), 11*
Moss, B., 177 (3, 7, 8, 9, 10), 179 (16), 180 (16), 181 (16), 190*
Moss, R., 177 (4), 190*
Mossige, J., 1 (21), 11*
Mühlbock, O., 1 (47), 3 (19, 20, 47), 4 (47), 11*, 12*
Müller, R., 140 (20), 146*, 196 (34), 204*
Munavalli, S. N., 194 (20), 204*
Munson, A. E., 74 (103), 76 (104), 84*
Muraoka, M., 193 (11), 194 (12, 13), 197 (12), 203*
Murayama, W., 165 (15), 175*
Murphy, E. B., 66 (74), 67 (75), 83*
Myers, D. D., 75 (92), 84*
Myrback, K., 166 (17), 175*
Myska, V., 140 (37), 146*

## N

Nagashima, N., 165 (15), 175*
Nagata, C., 171 (25), 175*
Nagayama, A. B., 177 (5), 190*
Naik, S. R., 200 (67, 68), 205*
Nakazato, H., 59 (15), 61*
Nandi, S., 3 (48), 12*
Naylor, R., 67 (6), 81*
Neauport-Sautes, C., 27 (17), 31*
Neidhardt, F. C., 45 (46), 62*
Neil, S. M., 166 (18), 170 (18), 171 (18), 175*
Nemes, M. M., 66 (50, 76, 94), 67 (52), 68 (78, 119), 73 (93), 74 (93), 75 (93), 77 (77), 82*, 83*, 84*, 85*, 195 (24), 204*
Ness, T. M., 66 (82), 83*
Neubauer, R. H., 75 (111), 84*
Neudeccker, T., 172 (26), 173 (26), 175*
Newcomb, J. M., 66 (10), 78 (10), 81*
Newmark, P., 194 (20), 204*
Newton, W. A., 119 (40), 132*
Ng, M. H., 70 (124), 77 (122, 124), 85*
Nicholson, M. D., 94 (17), 98*
Niderman, J. C., 27 (14), 31*
Nirenberg, M. W., 45 (27), 62*
Nishioka, K., 27 (16), 31*
Nissley, P., 60 (10), 61*
Noll, H., 101 (13), 102 (13), 113*
Nomura, M., 45 (3), 61*, 63*, 64 (66)
Nordlund, J. J., 77 (94a), 84*
Noronha, F., de, 18 (23, 24), 25*
Novelli, G. D., 45 (64), 63*, 199 (58), 205*
Nowinski, R. C., 1 (49), 2 (49), 7 (49), 12*
Nussenzweig, R., 74 (71), 83*
Nutt, R. F., 199 (54), 205*
Nuwer, M. R., 73 (34), 74 (34), 77 (33), 82*

## O

Obara, T., 99 (3), 113*
Oboshi, S., 17 (14), 25*
O'Brien, T. W., 101 (12), 113*
O'Connor, G. R., 74 (95), 84*
Oda, K., 35 (13), 37 (13), 44*
Odaka, T., 3 (50, 51), 12*
Oettgen, H. F., 27 (15), 31*
Ogata, K., 149 (13, 14, 15), 161*
Oh, J. O., 74 (95, 96), 78 (97), 84*
Oie, H., 78 (90), 84*
Okazaki, W., 3 (17), 11*
Oki, T., 149 (13, 14, 15), 161*
Old, L. J., 1 (49), 2 (49), 7 (49), 8 (58), 12*, 16 (15), 17 (1), 18 (4, 5), 21 (26), 24*, 25*, 27 (15), 31*
Olson, K. L., 99 (4), 113*
Omura, H., 142 (47), 146*
Oroszian, S., 18 (9), 25*
Orth, D. N., 2 (44), 11*
Ortwerth, B. J., 199 (58), 205*
Ostler, H. B., 74 (96), 78 (97), 84*
Ota, Z., 15 (10), 25*
Otis, H. K., 4 (23), 11*
Otter, A. B., 197 (43), 204*
Owens, R. B., 181 (25), 191*
Oxman, M. N., 59 (47), 60 (47), 62*

## P

Painter, R. B., 184 (38), 191*
Palese, P., 60 (20), 61*
Palsson, P. A., 120 (10, 11, 12), 131*
Pamplin, P. R., 66 (10), 81*
Panteleakis, P. N., 75 (81), 83*
Papaioannou, R., 168 (19), 169 (19), 170 (19), 171 (19), 172 (19), 175*
Paranchych, W., 93 (9), 97*
Pariser, R., 165 (11), 175*
Park, J. H., 73 (98), 74 (98), 84*
Parks, J. S., 60 (45, 48), 62*
Parks, W. P., 120 (41), 132*
Parr, R. G., 165 (11), 175*
Parsons, J. T., 104 (17), 113*
Pastan, I., 60 (10, 45, 48), 61*, 62*
Pasternak, G., 15 (2, 17), 16 (17, 20), 17 (16), 20 (17, 18), 21 (21, 22), 22 (13, 20), 23 (19), 24*, 25*
Pasternak, L., 16 (20), 20 (18), 21 (21, 22), 22 (20), 23 (19), 25*
Paucker, K., 47 (49), 62*

AUTHOR INDEX 219

Paymaster, J. C., 128 (35), 132*
Payne, L. N., 3 (52), 7 (52), 8 (53), 12*
Peacock, A. C., 103 (15), 110 (15), 113*
Pearson, G., 28 (9), 30*
Pegg, A. E., 200 (59), 205*
Peiffer, R. L., 191*
Pennington, T. H., 177 (6), 179 (13, 14, 15), 181 (13, 33), 182 (13, 14), 190*, 191*
Pereira, H. G., 87 (25), 98*
Perez-Bercoff, R., 77 (44), 82*
Perkins, F. T., 47 (4), 61*
Perkins, J. C., 74 (69), 83*
Perlman, R. L., 60 (45, 48), 62*
Persky, S., 149 (10), 152 (23, 24), 161*
Peschke, G., 172 (26), 173 (26), 175*
Peters, R. L., 18 (9), 25*
Peuman, S., 110 (22), 111 (22), 113*
Pflughaupt, K. W., 140 (20), 146*, 196 (34), 204*
Pichat, L., 197 (44), 198 (44), 204*
Piechowska, M., 194 (14), 195 (14), 196 (14), 203*
Pieroni, R. E., 78 (99), 84*
Pietrzykowska, I., 196 (32, 33), 204*
Pike, M. C., 139 (29), 146*
Pilch, D. J. F., 66 (101), 84*
Pina, M., 87 (10), 91 (19), 92 (18), 97*, 98*, 142 (24), 146*
Piraino, F., 3 (54), 12*
Pister, L., 18 (24), 25*
Pitha, J., 70 (100), 84*, 202 (75), 205*
Pitha, P. M., 70 (100), 84*, 202 (75), 205*
Planterose, D. N., 66 (101), 84*
Plevova, J., 140 (37), 146*
Pogo, B. G. T., 179 (12), 190*
Pogo, G. T., 177 (5), 190*
Pollack, Y., 45 (50, 51), 60 (50, 51), 62*
Pollard, M., 1 (47), 3 (47), 4 (47), 12*, 120 (12), 131*
Pope, J. H., 27 (16, 19), 31*
Pople, J. A., 165 (12), 175*
Postic, B., 74 (47, 48, 102), 82*, 84*

Priori, E. S., 119 (40), 132*
Prusoff, A. H., 140 (41), 145*
Prusoff, W. H., 135 (12), 136 (50), 137 (23, 49), 138 (28, 38), 139 (39, 58), 140 (23), 141 (42), 142 (1, 13, 14, 23, 39, 50), 143 (40), 144 (8, 45, 46), 145 (8a, 55a), 145*, 146*, 147*, 193 (1, 2), 194 (17), 197 (39), 198 (2), 199 (1), 203*, 204*
Pry, T. W., 112 (25), 113*
Prystas, M., 197 (42), 204*
Puchwein, G., 172 (26), 173 (26), 175*
Puliti, R., 149 (3), 161*
Pusztai, R., 94 (1), 97*

Q

Quastel, J. H., 194 (15), 204*
Quintrell, N., 104 (18), 113*

R

Rabbitts, T. H., 106 (20), 108 (20), 113*
Rabina, S., 150 (16), 161*
Rabotti, G. F., 2 (55), 12*, 112 (25), 113*
Rabson, A. S., 75 (85), 83*
Rada, B., 140 (15, 43), 146*
Raju, B. H., 139 (35), 146*
Rapp, F., 33 (4), 43*
Rappel, M., 140 (44), 146*
Raskova, H., 140 (37), 146*
Ray, R. K., 119 (39), 132*, 179 (18), 190*
Reamer, R. H., 3 (17), 11*
Reedman, B. M., 27 (16), 31*
Regamey, R. H., 47 (4), 61*
Regelson, W., 74 (103), 76 (104), 84*
Remington, J. S., 74 (105, 106), 84*
Renis, H. E., 74 (107), 84*
Revel, M., 45 (50, 51), 60 (50, 51), 62*
Rhim, J. S., 75 (108), 84*
Rhode, B., 193 (8), 203*

# AUTHOR INDEX

Rice, J. M., 77 (109), 84*
Rice, N. R., 77 (109), 84*
Riehm, E., 140 (19), 146*
Riggs, J. L., 87 (23), 98*
Riley, F., 76 (84), 83*
Riman, J., 99 (6, 7), 100 (7, 10), 103 (10), 104 (7), 112 (26), 113 (10, 26), 113*, 119 (36), 132*
Rita, G., 46 (41), 62*, 77 (39, 40), 82*
Robins, M. J., 200 (67, 68), 205*
Robins, R. K., 135 (12), 145*, 197 (39, 40, 41), 204*
Rochstel, M., 194 (15), 204*
Rokutanda, H., 119 (39), 132*
Rokutanda, M., 119 (39), 132*
Rosenbloom, F. M., 166 (16), 175*
Rosenblum, E. N., 177 (3, 4, 7, 8), 190*
Rosenfeld, C., 75 (91), 84*
Rossman, T. G., 77 (123), 85*
Rosztoczy, I., 78 (110), 84*
Roth, B., 193 (10), 203*
Rottman, F., 200 (60, 63), 202 (76), 205*
Roumiantzeff, M., 50 (52), 62*
Rowe, W. P., 35 (3), 43*
Roy-Burman, P., 193 (4), 199 (4), 203*
Rudland, P. S., 45 (14, 58), 61*, 63*
Rueckert, R. R., 47 (7), 50 (7), 61*
Rupp, W. D., 144 (45, 46), 146*
Russell, E. S., 1 (56), 12*
Russell, W. C., 87 (20), 93 (16), 98*, 142 (47), 146*

## S

Sachs, F., 136 (56), 139 (56), 147*
Sachs, L., 35 (1), 37 (1), 43*
Sakar, N. H., 1 (49), 2 (49), 7 (49), 12*
Sakouhi, M., 75 (91), 84*
Salb, J. M., 46 (40, 41), 62*
Salvi, M. L., 34 (10), 43*
Santesson, L., 27 (6), 30*
Saporoschetz, 4 (22), 11*
Sarid, S., 45 (11), 61*

Sarin, P. S., 119 (40), 132*
Sarker, N. H., 128 (35), 132*
Sarma, P. S., 18 (9), 25*, 75 (111), 84*
Sartorelli, A. C., 199 (53), 205*
Sather, G. E., 74 (102), 84*
Sauer, G., 33 (14), 34 (14), 35 (5, 14), 37 (16), 38 (14, 15, 16), 43*, 44*
Sawicki, L., 74 (18), 81*
Schabel, F. M., 140 (48), 146*
Schachter, J., 74 (96), 84*
Schäfer, W., 18 (23, 24), 25*
Schariff, M. D., 87 (15), 98*
Schedl, P. D., 45 (53), 62*
Scherberg, N. H., 45 (54, 65), 62*, 63*
Schildkraut, C. L., 92 (21), 98*
Schlesinger, R. W., 87 (22), 98*
Schlom, J., 7 (57), 12*, 115 (6, 7, 8), 116 (6), 117 (6), 118 (6), 119 (6, 7, 8, 25), 120 (20), 121 (7, 18, 20), 123 (20), 124 (20, 25), 125 (25), 128 (6, 34), 129 (8), 131*, 132*
Schmidt, F., 18 (24), 25*
Schmunis, G., 74 (18), 81*
Schnebli, H. P., 170 (23), 172 (23), 173 (23), 175*
Schneider, M., 75 (91), 84*
Scholtissek, C., 67 (56), 82*
Schramm, T., 15 (2), 24*
Schulte-Holthausen, H., 27 (6), 30*
Schwarzenberg, L., 75 (91), 84*
Scolnick, E., 119 (17), 121 (17), 131*
Scolnick, E. M., 119 (38), 120 (38, 41), 132*
Scott, W., 27 (16), 31*
Scriba, M., 28 (9), 30*
Seeiller, J. E., 166 (16), 175*
Seifert, E., 18 (24), 25*
Sekellick, M. J., 59 (42), 62*
Sekely, L., 136 (50), 137 (49), 141 (41), 142 (50), 146*, 194 (17), 204*
Senior, B. W., 45 (55), 63*
Seto, Y., 194 (13), 203*
Severi, L., 2 (55), 6 (10), 11*, 12*
Sezaki, T., 15 (10), 25*
Sharpe, T. J., 66 (101), 84*
Shedden, W. I. H., 142 (32), 146*
Sheffield, J. B., 3 (28), 11*

# AUTHOR INDEX

Sheldrick, P., 143 (53), 147*
Shen, T. Y., 198 (50), 204*
Shiff, W., 197 (38), 204*
Shimizu, Y., 165 (15), 175*
Shingu, H., 174 (29), 175*
Shiu, G., 75 (111), 84*
Shmueli, U., 149 (4), 161*
Shugar, D., 140 (52), 147*, 194 (14), 194 (14), 196 (14, 28, 29, 30, 31, 32, 33), 197 (31), 198 (46, 49), 200 (61, 62, 64, 65, 66), 201 (61, 64, 65), 202 (62, 78), 203 (79), 203* 204*, 205*
Sidikaro, J., 63*, 64 (66)
Sidwell, R. W., 198 (48), 204*
Siegert, W., 60 (20), 61*
Sierakowska, H., 200 (66), 205*
Sigurdsson, B., 120 (10, 11, 13), 131*
Silvestre, D., 27 (17), 31*
Silvestri, L., 179 (15), 190*
Simms, E. S., 126 (28), 132*
Simon, M. I., 70 (21), 81*
Sims, H. L., 75 (126), 85*
Singer, R. E., 45 (53), 62*
Sinsheimer, R. L., 122 (22), 131*
Sirsat, S. M., 128 (35), 132*
Skehel, J. J., 87 (20), 98*
Slechta, L., 194 (16), 204*
Smetana, K., 110 (21), 111 (21), 112 (21), 113*
Smith, A. E., 47 (56), 63*
Smith, D. W. E., 45 (60), 63*
Sonnabend, J. A., 45 (57), 46 (31, 57), 62*, 63*
Sorm, F., 197 (42), 204*
Soudry, E., 179 (17), 190*
Southam, C. M., 74 (64), 83*
Spiegelman, S., 7 (57), 12*, 115 (6, 7, 8), 116 (6), 117 (6, 9), 118 (6), 119 (6, 7, 8, 25), 120 (20), 121 (7, 18, 20), 122 (24), 123 (20), 124 (20, 25), 125 (25), 126 (27, 30), 127 (27, 30), 128 (6, 34), 129 (8), 131*, 132*
Spiess, G., 172 (26), 173 (26), 175*
Spirin, A. S., 101 (14), 102 (14), 104 (14), 108 (14), 109 (14), 113*
Srinivasan, P. R., 45 (63), 63*
Stancek, D., 47 (49), 62*
Steck, T. L., 100 (9), 104 (9), 113*

Steeves, R. A., 17 (25), 25*
Steinberg, C. M., 87 (5), 97*
Steitz, J. A., 45 (58), 63*
Stellwagen, R. H., 143 (51), 146*
Steplewski, Z., 37 (11), 43*
Stern, R., 67 (6), 77 (111a), 84*, 87*
Sternbach, H., 68 (35), 69 (35), 77 (35), 82*, 202 (77), 205*
Sternberg, S. S., 78 (134), 85*
Steuart, C. D., 199 (51), 204*
Stewart, R. C., 138 (17), 146*
Stewart, W. E., II, 78 (112), 84*
Stineberg, W. R., 137 (27), 146*
Stinebring, W. R., 71 (1), 77 (113), 78 (1), 81*, 84*
Stockert, E., 7 (58), 8 (58), 12*, 16 (15), 25*
Stoker, M. G. P., 45 (43), 62*
Stollar, B. D., 70 (21), 81*
Stoltzfus, C. M., 47 (7), 50 (7), 61*
Stone, H. A., 3 (17), 11*
Stone, L. B., 119 (17), 121 (17, 19), 131*
Stone, O. M., 66 (3), 71 (3), 81*
Stoops, C. E., 191*
Stück, B., 21 (26), 25*
Suarez, F., 35 (7), 43*
Subak-Sharpe, J. H., 177 (2), 179 (15), 190*
Subirana, J. A., 149 (4), 161*
Subrahmanyan, L., 27 (14), 31*
Sueoka, N., 45 (27, 59), 62*, 63*
Summers, D. F., 50 (52), 62*
Suskind, R. G., 112 (25), 113*
Sussman, M., 87 (5), 97*
Sutherland, E. S., 66 (10), 81*
Suwalsky, M., 149 (4), 161*
Suzaki, M., 15 (10), 25*
Svedmyr, A., 28 (18), 31*
Sverak, L., 99 (1, 2), 112 (2), 112 (26), 113 (26), 113*
Swann, P. F., 200 (59), 205*
Swetly, P., 42 (2), 43*
Swierkowski, M., 140 (52), 147*, 196 (28, 29, 30, 31), 197 (31), 204*
Szalay, G., 6 (10), 11*
Szer, W., 202 (78), 205*
Szilagyi, J. F., 177 (6), 179 (14, 15), 182 (14), 190*
Szybalski, W. H., 143 (53), 147*

## T

Tabor, C. W., 149 (1, 6, 11), 161*
Tabor, H., 149 (1, 6, 11), 161*
Takada, A., 193 (11), 194 (12), 197 (12), 203*
Takemori, M., 87 (23), 98*
Takemoto, K., 119 (17), 121 (17), 131*
Takemoto, K. I., 121 (19), 131*
Tan, Y. H., 77 (114, 115), 84*
Taube, A., 197 (43), 204*
Tavitian, A., 113 (27), 113*
Taylor, J. J., 78 (131), 85*
Temin, H. M., 7 (59, 60, 61, 63), 12*, 115 (1, 3, 4), 119 (4), 129 (3), 131*, 138 (54), 147*
Thiry, L., 179 (18), 182 (34, 35, 36), 190*, 191*
Thomas, C. A., Jr., 87 (10), 97*
Thomas, M. T., 83*
Thormar, F. H., 119 (16), 121 (16), 131*
Thormar, H., 120 (12, 14), 131*
Tillack, T. W., 45 (60), 63*
Tilles, J. G., 77 (116), 84*
Timbury, M. C., 177 (2), 179 (15), 190*
Timmermans, A., 2 (64), 4 (11), 11*, 12*
Ting, R. C., 180 (19), 190*
Tobin, J. O'H., 138 (3), 139 (3, 10), 145*
Tocchini-Valentini, G. P., 181 (28, 30, 31), 191*
Todaro, G. J., 3 (31), 6 (31), 7 (31), 11*, 20 (8), 25*, 38 (17), 44*, 119 (38), 120 (38, 41), 132*
Tomida, L., 149 (13, 14, 15), 161*
Tomkinks, G. M., 143 (51), 146*
Toy, S. T., 67 (88), 84*
Traub, W., 149 (4), 161*
Travers, A., 45 (61), 60 (61), 63*
Travnicek, M., 7 (57), 12*, 99 (5, 6), 113*, 115 (6), 116 (6), 117 (6), 118 (6), 119 (6, 25), 120 (20), 121 (20), 123 (20), 124 (20, 25), 125 (25), 128 (6), 131*
Trischmann, H., 201 (71), 205*
Trotter, J., 143 (9), 145*

Trown, P. W., 76 (87), 78 (87), 84*
Tsuda, K., 163 (3), 175*
Turner, H. C., 18 (9), 25*, 35 (3), 43*
Turner, W., 76 (117), 77 (109), 84*
Tushinski, R. J., 59 (12), 61*
Tyler, H. R., 139 (4), 145*
Tyler, J. M., 66 (3), 71 (3), 81*
Tytell, A. A., 66 (50, 76, 94, 118), 67 (52), 68 (49, 51, 78, 119), 73 (93), 74 (93), 75 (93), 77 (77), 82*, 83*, 84*, 85*

## U

Ueda, T., 193 (6, 7, 11), 194 (12, 13), 195 (7), 197 (12), 203*
Uhlendorf, C., 67 (11), 77 (11), 81*
Ulbricht, T. L. V., 197 (45), 204*
Uretzky, S. G., 113 (27), 113*
Ustacelebi, S., 87 (27), 88 (28, 29), 90 (29), 94 (24, 28), 98*

## V

Vaheri, A., 180 (23), 181 (26), 190*, 191*
Vaidya, A. B., 128 (35), 132*
Valentine, R. C., 87 (25), 98*
Van den Berghe, H., 78 (12), 81*
Vandeputte, M., 75 (120), 85*
Vanderberg, J., 74 (71), 83*
Vanderpool, E. A., 94 (7), 97*
Vanfrank, R. M., 66 (74), 83*
Van Kirk, J. E., 74 (69), 83*
Van Tiegham, N., 180 (22), 187 (22), 190*
Varacalli, F., 77 (123), 85*
Varmus, H. E., 60 (45), 62*
Vaughan, M., 59 (15), 61*
Vigier, P., 2 (65), 12*
Vilcek, J., 45 (62), 46 (62), 63*, 70 (124), 74 (59, 71), 75 (54), 77 (121, 122, 123, 124), 78 (5), 81*, 82*, 83*, 85*
Visser, D. W., 195 (23), 204*
Vitagliano, V., 149 (3), 161*
Vogt, P. K., 3 (66), 12*

# AUTHOR INDEX

Von Essen, C. F., 139 (6), 145*
Voytek, P., 145 (55a), 147*

## W

Wacker, A., 137 (11), 145*
Wahren, B., 21 (27), 25*
Wainfan, E., 45 (63), 63*
Waislren, B. A., 195 (27), 204*
Waitz, J. A., 74 (125), 85*
Walker, W. E., 195 (27), 204*
Wall, R., 59 (12), 61*
Walters, M. K., 27 (16, 19), 31*
Waters, L. C., 45 (64), 63*
Walton, E., 199 (54), 205*
Waltuck, G., 136 (56), 139 (56), 147*
Waterson, A. P., 87 (12), 97*
Watsen, D. K., 142 (32), 146*
Watson, K., 7 (57), 12*, 115 (6), 116 (6), 117 (6), 118 (6), 119 (6, 25), 120 (20), 121 (20), 123 (20), 124 (20, 25), 125 (25), 128 (6), 131*
Watson, K. F., 126 (27), 127 (27), 132*
Waubke, R., 28 (9), 30*
Weinberg, R. A., 110 (22), 111 (22), 113*
Weinstein, A. J., 75 (126), 85*
Weinstein, M. J., 74 (125), 85*
Weiss, S. B., 45 (21, 22, 34, 54, 65), 61*, 62*, 63*
Weissenbacher, M., 74 (18), 81*
Weissman, C., 60 (35), 62*
Weissmann, C., 181 (24), 190*
Welch, A. D., 137 (11), 138 (16, 28), 139 (6, 7, 57, 58), 145*, 146*, 147*
Wells, P. R., 164 (6), 175*
Wells, R. D., 69 (32, 36), 70 (32, 36, 37), 71 (37), 77 (32, 36), 82*
Wensink, P. C., 87 (10), 97*
Werner, G. H., 182 (37), 191*
Wheelock, E. F., 75 (127), 85*
White, D. O., 87 (15), 98*
Wildy, P., 87 (12), 97*, 142 (47), 146*

Willems, M., 110 (22), 111 (22), 113*
Williams, J. F., 87 (27), 88 (26, 28, 29), 90 (29), 94 (24, 28), 98*, 177 (2), 179 (15), 190*
Williams, L. B., 21 (3), 24*
Williamson, B., 27 (15), 31*
Wilson, D. H., 142 (47), 146*
Winchurch, R., 76 (128), 85*
Winkley, M. W., 197 (41), 204*
Winocour, E., 35 (1), 37 (1), 43*
Wood, H. A., 17 (1), 24*
Woodhour, A. F., 76 (129), 85*
Woodman, R. J., 198 (47), 204*
Wooles, W. R., 76 (104), 84*
Wolff, S. M., 77 (94a), 84*
Work, T. S., 47 (29, 30), 62*, 106 (20), 108 (20), 113*
Worthington, M., 73 (130), 74 (69), 74 (130), 83*, 85*

## Y

Yamada, H., 149 (13, 14, 15), 161*
Yamamoto, T., 3 (51), 12*
Yang, S. S., 180 (19), 190*
Yata, J., 27 (17), 29 (20), 31*
Yin, F. H., 67 (88), 84*
Yonezawa, T., 171 (25), 174 (27), 175*
Yoon, J. S., 9 (14), 11*
Young, P. A., 78 (131), 85*
Youngner, J. S., 77 (132, 133), 85*
Yu, M. C., 78 (131), 85*

## Z

Zajac, B. A., 28 (9), 30 (21), 30*, 31*
Zajdela, F., 75 (62), 83*
Zakay-Rones, Z., 181 (32), 191*
Zedeck, M. S., 78 (134), 85*
Zeleznick, L. D., 75 (135), 85*
Zelljadt, I., 128 (32, 33), 132*
Zischka, R., 99 (1), 113*
Zmudzka, B., 200 (61, 62, 64), 201 (61, 64), 202 (62), 205*

# Subject Index

## A

Acrolein, effect on NDV infectivity, 157
Actinomycin D, effect in interferon production, 77
Adenine analogues,
 binding affinities, 167
 electronic characteristics, 166
Adenine phosphoribosyltransferase,
 enzymic conversion rates, 171
 reaction rates, 170
 substrate requirements, 169
Adenosine kinase,
 binding properties, 170
 substrate conversion rates, 172-174
Adenosine kinase substrates,
 conversion rates, 172
 pi-electronic characteristics, 173-174
Adenovirus type 5 $ts$ mutants,
 antigen synthesis, 93
 DNA synthesis, 91-92
 genetic studies, 88-90, 92
 interferon induction, 94, 96
 isolation, 87
 physiology, 90-93
 transformation of cells, 94
Adenovirus type 12-transformed cells,
 effect of aminopiperazine on liquid metabolism, 189
 effect of rifamycin derivatives, 180
Aldehydes, antiviral activity, 156-157
Aminopiperazines,
 effect on DNA polymerase activity *in vitro*, 186, 192
 Epstein-Barr virus growth, 182
 herpes virus growth, 182-183, 189
 lipid metabolism of cells, 189
 myxovirus growth, 182

Aminopiperazines—*cont.*
 plaque formation by influenza virus, 187, 191
 thymidine incorporation into cells, 183-184
 transformed cells, 187-188
 vaccinia virus growth, 181-183
Antigens, viral,
 expressions of, by activation, 23-24
 Graffi leukaemia virus, 20, 22
 Gross leukaemia virus, 21-22
 herpes virus, 27
 murine leukaemia virus, 16-19, 20, 23
 of Epstein-Barr virus, 28, 30
Antigenic changes,
 in cells infected with RNA tumour viruses, 15-24
 in cells infected after transformation, 21-23
 in cells transformed by RNA tumour viruses, 16-20
 in non-malignant cells infected with RNA tumour viruses, 20-21
Avian myeloblastosis virus,
 ATPase activity, 26
 DNA-DNA polymerase, 122-125
 preparation of labelled virus and RNA, 99
 presence of non-viral RNA, 99
 purification of DNA polymerase, 127
 reverse transcriptase, 116, 119, 121, 126-127
Avian tumour viruses, host susceptibility, 3
6-Azathymidine, effect on herpes simplex virus growth, 135
6-Azauridine, antiviral activity, 140

## B

Bromo-deoxyuridine,
  effect on enzymic formation in mammalian cells, 143

## C

Colicin, as inhibitor of protein synthesis, 45
Cycloheximide, effect on interferon production, 77
Cytosine arabinoside, activiral activity, 139

## D

DNA, as interferon inducer, 65-66, 70
  binding to oxidized spermine, 151
  effect on reverse transcriptase reactions, 122
  template for oncornavirus polymerases, 124
DNA tumour viruses (See also S.V. 40 virus)
  DNA synthesis and mRNA transcription, 33
  DNA synthesis and transformation, 38-43
  expression of genome, 33-34
  relationship to DNA replication, 33-34
  transcription and protein synthesis, 34-36
  transcription in transformed cells, 37-38

## E

Encephalomyocarditis virus,
  general properties, 47
  polypeptide synthesis in vitro, 48-51
  polypeptide synthesis, effect of interferon, 53-55
  translation of RNA genome, 50, 54
Epstein-Barr virus,
  antigens induced, 27-30

Epstein-Barr virus—cont.
  effect of aminopiperazines on growth, 182
  general properties, 27
  macromolecular synthesis in infected cells, 29-30
  possible oncogenicity, 27
  susceptibility of cell lines to infection, 29

## F

Formaldehyde,
  effect on influenza virus haemagglutinin, 158
  Newcastle disease virus infectivity, 157
  vaccinia virus infectivity, 157
Freund's adjuvant, effect on interferon induction, 77

## G

Glutaraldehyde,
  effect on influenza virus haemagglutination, 158
  Newcastle disease virus infectivity, 157
  vaccinia virus infectivity, 157
Graffi leukaemia virus,
  antigen distribution in cells infected with, 20-22
  antigen suppression in Gross leukaemia cells, 21-22
Gross leukaemia virus,
  antigen expression, 21, 23
  interference with Graffi antigen expression, 21-22
Guanine analogues,
  electronic characteristics, 168
  enzymic conversion rates, 171

## H

Herpes simplex virus,
  effect of aminopiperazines on growth, 183, 189

Herpes simplex virus—*cont.*
 effect of 6-azathymidine on growth, 135
 effect of cytosine arabinoside on growth, 199
 effect of 5,6-dichlorobenzimidazole on growth, 135
 effect of 5-iodo-2-deoxyuridine on growth, 137, 142
 effect of N-methyl iododexoxyuridine on growth, 136
 nucleoside analogues, 195
 pyrimidine derivatives, 193-194
Herpes virus infections,
 therapeutic effect of 6-azauridine, 140
 therapeutic effect of cytosine arabinoside, 140
 therapeutic effect of iododeoxyuridine, 138-140
 therapeutic effect of polynucleotides, 73
Human milk, virus-like particles isolated from, 128-129
Hypoxanthine analogues,
 electronic characteristics, 168
 enzymic conversion rates, 171
Hypoxanthine phosphoribosyltransferase,
 reaction rates, 170
 substrate requirements, 168

I

Influenza virus (*See also* Myxoviruses),
 binding by oxidized spermine, 158
 inactivation by amino aldehydes, 153-156
 inactivation by oxidized spermine, 150, 153-156, 158-160
 plaque number, effect of aminopiperazine, 182-183, 191
 plaque size, effect of aminopiperazine, 191
Influenza virus haemagglutination,
 effect of formaldehyde, 158
 effect of glutaraldehyde, 158
 effect of oxidized spermine, 158

Interferon,
 antitumour activity, 75-76
 effect on EMC virus polypeptide synthesis, 50
 effect on interferon induction, 74-75, 78-80
 effect on microorganisms, 74
 effect on S.V. 40 in RNA synthesis, 59
 effect on vaccinia virus in RNA synthesis, 56, 59
 effect on V.S.V. in RNA synthesis, 59
 effect on viral protein synthesis, 45
 mechanism of action, 46, 55, 60
 purification, 64
 relationship with repressor content of provirus, 12
Interferon induction,
 by adenovirus type 5, 94-96
 by DNA-RNA hybrids, 70
 by DNA viruses, 67
 by nucleic acids, 65-81
 by RNA viruses, 66-67
 by polydeoxyribonucleotides, 70
 by polyribonucleotides, 69, 70
 effect of alkylated polynucleotides, 202-203
 effect of DNA polymers, 71
 effect of Freund's adjuvant, 77
 effect of lead acetate, 77
 effect of metabolic inhibitors, 77
 polycations, 77
 potentiation of interferon response, 76-78
 structural requirements of inducer, 68-70
 substituted polyribonucleotides, 77
 synthetic polyribonucleotides, 68, 71-73
 the triggering molecule, 66-67
 the triggering process, 70-73
5-Iodo-2'-deoxyuridine,
 antiviral activity, 138, 142-143
 as therapeutic agent, 138-140
 base pair errors caused by, 144
 crystalline and molecular structure, 143
 effect on cellular enzymes, 142
 effect on herpes simplex virus growth, 137, 142

# SUBJECT INDEX

5-Iodo-2'-deoxyuridine—*cont.*
  incorporation into DNA, 142-143, 147
  interaction with virus and host cells, 134-144
  mechanism of action, 141
  metabolism, 137
  pH effect on activity, 148
  photochemical studies, 147
  pseudorabies virus growth, 142
  Rous sarcoma virus growth, 138
  sensitization of enzymes to radiation in activation, 144
  toxicity, 137-138
  uninfected cells, 137
  vaccinia virus growth, 142-143

## L

Leukaemia,
  activation by carcinogens, 15
  induction of disease, 15
  susceptibility to infection, 15
  tumour-associated antigens, 15

## M

Mason-Pfizer monkey virus,
  DNA polymerase activity, 128
  general properties, 128
Mouse mammary tumour virus,
  genetic transmission, 4
  host susceptibility, 3
  induction, 6
  partial expression, 7
  recovery of infectivity from tissues, 9
  reverse transcriptase activity, 118-121
  templates for reverse transcriptase, 124-125
  transcription, 6
Murine leukaemia,
  immunization against, 16
  transmission of diseases, 4
Murine leukaemia virus (*See also* Graffi and Gross leukaemia virus),
  antigens, 16-20, 23
  host susceptibility, 3
  induction, 4, 23

Murine leukaemia virus—*cont.*
  reverse transcriptase activity, 118-121
  templates for reverse transcriptase, 124-125
  streptovaricin and reverse transcriptase, 180
Myxoviruses,
  effect of aminopiperazine, 182
  effect of oxidized spermine and spermidine, 153-157
  EM appearance, after oxidized spermine treatment, 159

## N

Newcastle disease virus,
  cell fusion activity,
    effect of oxidized spermine, 159
  infectivity,
    effect of acrolein, 157
    effect of formaldehyde, 157
    effect of glutaraldehyde, 157
    effect of oxidized spermidine, 157
    effect of oxidized spermine, 157
Nucleoside analogues, antiviral activity, 195-201

## O

Oncornaviruses,
  antigenic changes in infected cells, 15-16, 20-23
  antigens of transformed cells, 16-20
  cellular genetic changes, 9
  general properties, 1
  genetic susceptibility, 2-3
  genetic transmission, 2-7
  incidence, 2
  induction by radiation and chemicals, 2
  influence of epigenetic state of host cells, 9-10
  integration into host cell genome, 7-8
  interaction with host cell genome, 1-2

# SUBJECT INDEX

Oncornaviruses—*cont.*
  neoplastic transformation, 9
  partial expression, 7
  replication, 12-13
  ribosomal components, 99-113
Oncornavirus reverse transcriptase,
  general properties, 7, 115-131
  inhibition by rifamycin derivatives, 179-180
  inhibition by streptovaricin, 180
  template requirements, 125
Oxidized polyamines,
  antimicrobial activity, 148-150
  antiviral activity, 148-162
  mode of action, 150-152
Oxidized spermidine, effect on myxoviruses, 153-157
Oxidized spermine,
  antiviral activity, 152-159
  binding to DNA, 151, 162
  binding to influenza virus, 158
  biological properties of viruses inactivated by, 157-158
  effect on influenza virus haemagglutination, 158
  effect on myxoviruses, 153-157, 159
  effect on NDV cell fusion activity, 159
  effect on NDV infectivity, 157
  effect on vaccinia virus infectivity, 157
  physical and biological effects, 152
  therapeutic effect, 158-159
  toxicity, 158-160

## P

Polydeoxynucleotides, interferon-inducing ability, 70
Polynucleotides,
  cell interaction and interferon production, 70
  molecular weight and antiviral activity, 86
  RNase resistance and antiviral activity, 86
  thermal activation and antiviral activity, 85

Polynucleotides—*cont.*
  toxicity and antiviral activity, 78-80
Polynucleotides,
  alkylated, antiviral activity, 201-203
  interferon-inducing ability, 202-203
  synthetic, interferon-inducing ability, 68
Polyribonucleotides, interferon-inducing ability, 69
Polyribonucleotides substitutes, interferon-inducing ability, 77
Polyribonucleotides, synthetic,
  antitumour activity, 75-76
  as template for oncornavirus polymerases, 124
  effect on interferon production, 73
  interaction with cells and interferon production, 71
  prophylactic effects, 73-74
Pyrimidine analogues, antiviral activity, 193-195
Pyrimidine nucleoside analogues, antiviral activity, 195-201
Protein synthesis,
  in cell face systems, 47-55
  in cells infected with DNA tumour viruses, 35-36
  in EMC-infected cells, 47-55
  in vaccinia virus-infected cells, 55-61
  in virus infected and interferon treated cells, 45-61
Pseudorabies virus, effect of 5-iodo-2'-deoxyuridine, 142
Purine metabolic pathways, molecular orbital study, 163-174
Purine phosphoribosyltransferase,
  properties, 166-170
  reaction rates, 170-174
  substrate requirements, 166

## R

Rauscher leukaemia virus,
  enhancement of leukaemic response, 9-10
  induction of erythroblastosis, 9
  reverse transcriptase activity, 119, 121

Rauscher leukaemia virus—*cont.*
  reverse transcriptase product, 117-118
  response to various templates, 124-125
Reverse transcriptase,
  activity in virus-like particles from human milk, 128-131
  DNA template, 122
  distinction from other DNA polymerases, 126-127
  hybridization studies on product, 117-118
  nature of product, 116-118
  of avian myeloblastosis virus, 126-127
  of Mason-Pfizer monkey virus, 128
Reverse transcriptase,
  of oncornaviruses, 7, 115-131
  of Visna virus, 121
  presence in RNA viruses, 118-121, 128
  properties of reaction, 115-116
  requirements for *in vitro* activity, 116
  rifamycin derivatives, effect on, 179
  synthetic templates, 124-126
RNA-dependent DNA polymerase (*See* reverse transcriptase)
Ribonucleic acid, double stranded,
  as interferon inducer, 65-67
  effect on host cell defence mechanisms, 76
  thermal stability and interferon induction, 68
Ribonucleic acid, ribosomal,
  analysis, 108-112
  electrophoresis, 103
  isolation, 103
Ribosomes, as components of oncornaviruses, 99-113
Ribosomes, cytoplasmic,
  co-sedimentation with virus ribosomes, 106-108
  isolation, 102
Ribosomes, viral,
  basic properties, 108-112
  co-sedimentation with cytoplasmic ribosomes, 106-108
  labelling and isolation, 99-101, 103

Ribosomes, viral—*cont.*
  labelling kinetics, 104
  sedimentation characteristics, 102
Rifampicin,
  effect on cell transformation, 181
  RNA bacteriophage growth, 179
  vaccinia virus growth, 177-179
Rifamycin derivatives,
  chemical formulae, 178
  effect on animal cells, 180-181
  effect on cell transformation, 180-181
  effect on intracellular synthesis, 181
  effect on microbial polymerases, 179
  effect on mitochondria, 180
  effect on reverse transcriptase reactions, 179-180
  effect on vaccinia virus growth, 177-179
Rous sarcoma virus,
  inhibition by 5-iodo-2'-deoxyuridine, 138
  mutagenicity, 9
  reverse transcriptase activity, 119, 121
  reverse transcriptase product, 117
  rifamycin derivatives, effect on growth, 181
  sarcomas, effect of polynucleotides, 75

**S**

Streptovaricins,
  association of DNA with cell components, 40
  association between viral and cellular DNA synthesis, 41-42
  characteristics of transformed cells, 37
  DNA synthesis and transformation of host cells, 38-43
  effect on DNA to host cells, 39
  effect of rifamycin derivatives on transformed cells, 180
  production of viral proteins, 35
  size of messenger RNA, 44
  structural changes in DNA, 38
  transcription of genome, 34-38

## SUBJECT INDEX

### T

Transformed cells,
  antigenic changes, 15-24
  DNA synthesis, 38-43
  effect of
    aminopiperazines, 187-189
    rifampicin, 181
    rifamycin derivatives, 180
  genetic changes, 9
  induced by
    adenovirus type 5 *ts* mutants, 94
    S.V. 40 virus, 37
  transcription, 37-38

### V

Vaccinia virus growth,
  effect of
    aminopiperazines, 181-183
    5-iodo-2'-deoxyuridine, 142-143

Vaccinia virus growth—*cont.*
  effect of—*cont.*
    rifampicin, 177
    rifamycin, derivatives, 177-179
  infectivity
    effect of
      formaldehyde, 157
      glutaraldehyde, 157
      oxidized spermine, 157
  macromolecular synthesis in infected cells, 55
  messenger RNA
    association with cellular proteins, 57
    effect of interferon, 59
    sedimentation profile, 56
    synthesis, 56
    structure, 56
Visna virus,
  general properties, 120
  nuclei acid, 121
  reverse transcriptase, 121